THE
CHEMOKINE
FactsBook

THE CHEMOKINE
FactsBook

Krishna Vaddi
Genzyme Corporation
Framingham, MA

Margaret Keller
duPont Hospital for Children
Wilmington, DE

Robert C. Newton
The DuPont Merck Pharmaceutical Co
Wilmington, DE

Academic Press
Harcourt Brace & Company, Publishers
SAN DIEGO LONDON BOSTON NEW YORK
SYDNEY TOKYO TORONTO

Academic Press, Inc.
525 B Street, Suite 1900, San Diego, California 92101-4495, USA
http://www.apnet.com

Academic Press Limited
24–28 Oval Road, London NW1 7DX, UK
http://www.hbuk.co.uk/ap/

ISBN 0-12-709905-0

Library of Congress Cataloging-in-Publication Data

Vaddi, Kris.
    The chemokine factsbook / by Kris Vaddi, Margaret Keller, Robert
Newton.
        p.   cm.
    ISBN 0-12-709905-0 (alk. paper)
    1. Chemokines—Handbooks, manuals, etc.   I. Keller, Margaret.
II. Newton, Robert, 1952– .   III. Title.
    [DNLM: 1. Chemokines—handbooks.   QW 539 V122c 1996]
QR185.8.C45V33   1996
616.07′9—dc21
DNLM/DLC
for Library of Congress   96-44883
                                                                                CIP

A catalogue record for this book is available from the British Library

Typeset in Great Britain by Alden, Oxford, Didcot and Northampton

For information on all Academic Press publications
visit our website at books.elsevier.com

Transferred to digital print 2008
Printed and bound by CPI Antony Rowe, Eastbourne

Contents

Section III THE CHEMOKINE RECEPTORS

Preface

We would like to thank Tessa Picknett and Academic Press for their patience and encouragement during the compilation of this work. We would also like to thank our spouses, Lakshmi, Grant and Toni, for their understanding and support. Much appreciation goes to Dr David W. Martin, founder of the DuPont Merck Postdoctoral program which brought the authors together and supported our interest in chemokine biology.

Left to right: *Robert Newton, Margaret Keller, Krishna Vaddi*

Abbreviations

ADP	Adrenaline activated platelets
AIDS	Acquired immune deficiency syndrome
AML	Acute myeloid leukemia
AP-1	Activator protein-1
ARDS	Adult respiratory distress syndrome
BFU	Burst-forming unit
BLR-1	Burkitt's lymphoma receptor 1
β-TG	Beta thromboglobulin
C/EBP	CCAAT/enhancer-binding protein
C5a	C5 anaphylatoxin
C5aR	C5 anaphylatoxin receptor
cAMP	cyclic adenosine monophosphate
CC CKR	CC chemokine receptor
CEF	Chick embryo fibroblasts
CFU	Colony-forming unit
CIDNP	Chemically-induced dynamic nuclear polarization
CINC	Cytokine-induced neutrophil chemoattractant
CK	Chemokine receptors
CMK-BRL-1	Chemokine beta receptor-like 1
CMV	Cytomegalovirus
Con A	Concanavalin A
CP	Capsular polysaccharide
CTAP-III	Connective-tissue activating peptide III
DARC	Duffy antigen/chemokine receptor
DMSO	Dimethylsulfoxide
DTH	Delayed type hypersensitivity
EAE	Experimental autoimmune encephalomyelitis
EBI-1	Epstein–Barr induced protein 1
EC_{50}	50% effective concentration
ECM	Extracellular matrix
ECRF3	An open reading frame of Herpes virus Saimiri gene sequence
EDTA	Ethylene diamine tetraacetic acid
ELISA	Enzyme-linked immunoabsorbent assay
ELR	Glutamic acid–leucine–arginine motif
EMF-1	Embryo fibroblast protein 1
ENA-78	Epithelial derived neutrophil attractant 78
ERK-1	Extracellular signal regulated kinase 1
ESP MS	Electron spray mass spectroscopy
FGF	Fibroblast growth factor
FITC	Fluoroscein isothiocyanate
fMLP	Formyl-methionyl-leucinyl-phenylalanine
FMLPR	*N*-formyl-methionyl-leucyl-phenylalanine receptor

FPR	*N*-formyl peptide receptor
FSH	Follicle stimulating hormone
G0, G1	Growth phase of the cell cycle
GBM	Glomerular basement membrane
GCP-2	Granulocyte chemotactic protein 2
GCSF	Granulocyte colony stimulating factor
GMCSF	Granulocyte-macrophage colony stimulating factor
GMP	Guanidine monophosphate
GMP-PNP	Non-hydrolyzable analog of guanidine monophosphate
GPCR	G-protein coupled receptors
GRE	Glucocorticoid response element
GRO	Growth related gene
GTP	Guanidine triphosphate
GTP-γS	Guanidine triphosphate (thioester)
H-RS	Hodgkin and Reed-Sternberg cells
HBV-X	Hepatitis B virus X protein
HCC-1	Human CC chemokine
HEK	Human embryonic kidney cells
HEL	Human erythro leukemia cells
HIV	Human immunodeficiency virus
HIV-SF	Human immunodeficiency virus suppressive factors
HPLC	High-pressure liquid chromatography
HSPG	Heparan sulfate proteoglycans
HTLV	Human T lymphoma virus
HUMSTSR	Human seven transmembrane segment receptor
HUVECs	Human umbilical vein endothelial cells
IC_{50}	50% inhibitory concentration
IFN	Interferon
Ig	Immunoglobulin
IL	Interleukin
IL-8RA	Interleukin 8 receptor A
IL-8RB	Interleukin 8 receptor B
IP-10	Gamma interferon inducible protein 10
IP_3	Inositol triphosphate
ISRE	IFN-stimulated response element
KC	Murine immediate early gene
KS	Karposi's sarcoma
LA-PF4	Low-affinity platelet factor 4
LAM	Lipoarabinomannan
LAM-1	Leukocyte adhesion molecule 1
LECAM-1	Leukocyte endothelial cell adhesion molecule 1
LESTR	Leukocyte-derived seven transmembrane domain receptor
LFA-1	Leukocyte function antigen 1
LH	Luteinizing hormone
LIF	Leukemia inhibitory factor
LPS	Lipopolysaccharide
LTB_4	Leukotriene B4
Ltn	Lymphotactin
MAC-1	Macrophage specific adhesion molecule

mCFU	Mixed colony-forming unit
MCP	Monocyte chemoattractant protein
MDR-15	Monocyte-derived receptor 15
MGSA	Melanocyte growth stimulatory activity
MIG	Monokine induced by IFN-γ
MIP	Macrophage inflammatory protein
MK	Megakaryocytes
MM-LDL	Minimally modified low density lipoprotein
mRNA	Messenger RNA
NADPH	Nicotinamide adenine dinucleotide phosphate
NAP-2	Neutrophil activating protein 2
NECA	N-ethylcarboxamidoadenosine
NF-κB	Nuclear factor κB
NF-IL6	Nuclear factor of interleukin 6
NK	Natural killer cell
NMR	Nuclear magnetic resonance spectroscopy
ORF	Open reading frame
PAF	Platelet activating factor
PAFR	Platelet activating factor receptor
PBMC	Peripheral blood mononuclear cells
PBP	Platelet basic protein
PDGF	Platelet derived growth factor
PF4	Platelet factor 4
PGE$_2$	Prostaglandin
PGI$_2$	Prostacyclin
PHA	Phytohemagglutinin A
PI	Inorganic phosphate
PIP$_2$	Phosphatidylinositol diphosphate
PKA	Protein kinase A
PKC	Protein kinase C
PLA$_2$	Phospholipase A$_2$
PLC	Phospholipase C
PMA	Phorbol myristate acetate
PMN	Polymorphonuclear leukocytes
PMSF	Phenyl methyl sulfonic acid
PPD	Purified protein derivative of tuberculin
PRH	Prolactin releasing hormone
PT	Pertussis toxin
RA	Rheumatoid arthritis
RANTES	Regulated on activation, normal T expressed and secreted
RPE	Retinal pigment epithelial cells
RSV	Rous sarcoma virus
RT–PCR	Reverse transcriptase–polymerase chain reaction
SCID	Severe combined immunodeficiency
SCM-1	Single cysteine motif-1
SDF-1	Stromal cell derived factor-1
SDS–PAGE	Sodium dodecyl sulfate–polyacrylamide gel electrophoresis
SF	Synovial fluid

SIG	Small inducible gene
SIV	Simian immunodeficiency virus
SLF	Steel factor
SRBC	Sheep red blood cell
SRU	*src*-responsive unit
STAT-1	Signal transducer and activator of transcription-1
TF	Transcription factor
TIL	Tumor-infiltrating lymphocyte
TNF	Tumor necrosis factor
TPA	Tumor promoting acetate

THE INTRODUCTORY CHAPTERS

Section 1

THE
INTRODUCTORY
CHAPTERS

1 Introduction

AIM OF THE BOOK

The primary goal of this book is to provide researchers with a source of comprehensive information on the rapidly evolving family of intercellular mediators known as 'chemokines'. As new proteins of biological importance are being identified almost every day, it is nearly impossible to keep up with their functions and other facts that make them useful tools in our daily research. These proteins are exquisitely specific in their effects on cell types and the responses they elicit. We often find ourselves digging through the literature to find what they do, how they are regulated and how they could be mediating the effects we find in our own research. The authors of the previously published *Cytokine FactsBook* did a phenomenal job of putting together most of the facts on cytokines and included extensive information on chemokines as well. We have tried to make the present book complementary to the *Cytokine FactsBook* by providing more detailed information on the biological effects, regulation of expression, expression in different disease states (both human diseases and animal models), in addition to the basic facts that readers have come to expect from a book in this series.

HOW DO CHEMOKINES DIFFER FROM CYTOKINES?

Although still considered as members of the cytokine superfamily, chemokines are rapidly establishing their own identity as a class of molecules with very distinct effects. The name chemokine comes from a combination of *chemo*tactic and cyto*kines* and chemotaxis or signaling for directed migration has been the central concept, besides the structural similarities, that distinguishes this class of proteins as a family. It is clearly established that many pathological as well as physiological processes are closely regulated by a given cell type, and scientists have long wondered about the signals that could specifically draw a given cell into a given tissue. As Schall aptly put it 'Like the physicists' dark matter, the existence of chemokines had been long suspected, but their nature in the immunological cosmos was undefined until recently' [1].

With the identification of IL-8, which is chemotactic to granulocytes but not monocytes, it appeared clearly possible to have a protein that could specifically deliver a migratory signal to a specific cell type. This finding was in sharp contrast to those that show lack of cell specificity such as formyl peptides, C5a, and LTB$_4$. Discovery of MCP-1 represents another important milestone in the evolution of the chemokine family. MCP-1 is primarily chemotactic to mononuclear leukocytes but not polymorphs, and hence seen to play an important role in chronic inflammatory diseases. Identification of IL-8 and MCP-1 marks a new era in chemotaxis biology, and serious efforts have begun to explore other proteins such as those that could be chemotactic to other cell types.

Another important turning point for the chemokine field came about with the identification of heptahelical receptors for IL-8. Subsequently several chemokine receptors have been identified, and as far as we know today, chemokines are the only members of the cytokine superfamily that bind to receptors that are seven-transmembrane and G-protein coupled in nature; another feature that makes the chemokines distinct. These findings also opened up a whole new area of research exploring new chemokine receptors and how each of the chemokines interacts with their receptors. Intense research is currently focused on understanding the specificity of chemokines and their receptors utilizing structural and functional studies with the goal of identifying of molecules that could specifically modulate receptor–ligand interactions.

However, as of today, we are not really sure if the most important physiological function of chemokines is chemotaxis alone. Recent findings[2] on the molecular basis of the angiogenic and the angiostatic effects of chemokines have sparked a great interest in their role in physiological and pathological conditions that require neovascularization. Even more recently, the discovery of a relationship between β-chemokines and the resistance to human immunodeficiency virus infection[3] has created newer avenues of scientific exploration of the real biological role of chemokines. Other novel biological effects of chemokines such as T-cell costimulatory effects are also beginning to emerge[4].

This book has been organized to present the current state of knowledge. Only the published literature has been included since it is accessible to all parties and represents the body of knowledge which has been reviewed and discussed by the research community. The results sections of most papers were reviewed and used to locate additional information wherever possible. For example, a paper on the expression of a particular chemokine may also include relevant data on the biological activities of the protein. In another instance a paper on the *in vivo* role of a particular chemokine may also include important data generated during the quality control of the protein.

Although every effort has been made to include the major properties of each chemokine, there are undoubtedly a number of omissions. Whereas chemotaxis has been in the literature since before the turn of the century, the chemokine field is a relatively young one and most references included in this review are from 1986 or later. Over 1000 references have been selected, reviewed and included in this summary with over 80% of those from the years 1993–95. Since the field continues to evolve and new literature appears almost daily, this body of knowledge needs to become a living document. Any reader who is interested in the latest information or in particular references for any inclusion is invited to contact the authors.

References
[1] Schall, T. (1994) The chemokines. In *The cytokine handbook* (ed. A.Thomson). Academic Press, London.
[2] Koch, A.E. et al. (1992) *Science* 258, 1798.
[3] Cocchi, F. et al. (1995) *Science* 270, 1811.
[4] Taub, D.D. et al. (1996) *J. Leuko. Biol.* 59, 81–89.

2 Organization of the data

The type of data presented differs between chemokine and chemokine receptor entries, and hence they are organized with a different structure.

CHEMOKINE ENTRIES

Chemokine entries include the following information:

Alternate names

This section lists names which appear in the literature and can be directly related to one of the proteins discussed. As in other fields, many proteins were cloned and identified by an alphanumeric code until their biological identity or homology to known proteins from either the same or other species was revealed. Attempts have been made to designate when the name was applied to a human or animal protein.

Family

Chemokines described in this book belong to three families, known as the α, β, and γ or CXC, CC, and C families.

Molecule

This section gives a general flavor of the entry by mentioning some salient features of the given chemokine.

Tissue sources

A list of tissues or cell types that produce or express the protein or gene is included. Some of the cell types require external stimulus to produce the chemokines, whereas some cells and tissues express constitutively.

Target cells

Cellular specificity is one of the distinguishing features of the chemokine family. A list of cell types that respond to a given chemokine is included in each entry.

Physicochemical properties

This section includes data on the molecular weights of the chemokines, number of amino acids, length of signal peptides, length of mature peptides, number and location of putative N-glycosylation sites, and number and location of disulfide bonds.

Transcription factors

In recent years a significant effort has been mounted to modulate the transcription factors (TF) therapeutically. We have included information about the promoters, sites where known TFs bind, and potential binding sites for different TFs with each chemokine entry. This information is thought to be gaining more importance as we understand the subtle differences in the transcriptional control of the chemokines.

Regulation of expression

Usually the earliest studies following the identification of a new protein involve an *in vitro* examination of the various cell types which produce the protein. This section is categorized in two ways. Where an extensive body of information is available, we have organized the data into a tabular form with the cell type, stimulus, and response. In other cases the information is included by cell type, with tumor cell lines generally located with their associated cell type. Within a cell type, studies are grouped by species of cell type. In general, where mRNA is mentioned, detection was by northern blot or, in more recent studies, by RT PCR. Protein is detected by Western blot analysis, by immunoprecipitation, by antibody neutralization, by immunohisto-chemistry, or by ELISA. For brevity, any apparently nonspecific reference to mRNA or protein within a section refers not to total mRNA or protein but to the specific chemokine under discussion. If the reference is to total mRNA or protein this is specifically stated.

Expression in disease

This section is further divided into human and animal studies. Since the role of any cytokine in human disease is of particular interest, this section focuses on studies in humans or in human biopsies which look for expression of chemokines. In most studies, mRNA is detected by RT PCR of tissue extracts or by *in situ* hybridization. Protein is detected by immunohistochemistry or by ELISA of tissue homogenates. Studies are generally grouped by tissue type studied.

Many animal models of human disease have been developed and this section focuses on studies which are analogous to those presented in the section on human disease expression. As in those studies, mRNA is detected by RT PCR on tissue extracts or by *in situ* hybridization. Protein is detected in most cases by immunohistochemistry. Studies are generally grouped by tissue type studied.

In vivo studies

This section contains studies where the protein was administered to animals or in some cases humans, either to measure specific migration of a cell population or to examine the effects on a biological process.

In vitro biological effects

This section is a very abbreviated listing of the major biological effects which result from the exposure of particular cells to a chemokine. Active concentrations for the biological effects are provided in parenthesis. The listing is organized by cell type and lists the types of outcomes measured with that cell type. Since chemokines in many instances exhibit a bell-shaped dose–response curve, a general dose range reported to be active in different studies is included. The reader is referred to the literature for common methodologies. For example, virtually all chemotaxis studies utilize the Boyden chamber methodology and cell-free calcium is studied through the use of calcium sensitive fluorescent dyes.

Intracellular signaling

This section includes the second messengers involved in the elicitation of biological responses to the chemokines.

Receptor binding characteristics

This section is treated in two or three different ways depending on the availability of the data. In most cases both radioligands and competing ligands have been noted since several researchers note significant differences in the binding affinities depending on how the experiments are conducted. Information on mutant peptides is also included here.

Cross-desensitization

Cross-desensitization is a well-characterized method to determine receptor sharing and specificity. By sequential challenge with the same chemokine (also known as homologous desensitization) or different chemokines (heterologous desensitization) one can find out if the two chemokines in question are interacting with the same or different receptors on the cell surface.

Gene structure

The gene structure includes information on the number of introns and exons and how they are structured. In most cases a graphic representation of the gene is included. The filled boxes indicate untranslated exons and open boxes are translated exons.

Gene location

Chromosomal localization of genes for human and murine chemokines is included in most cases.

Protein structure

Structural information in most cases has come from three-dimensional NMR or X-ray crystallographic studies. In some cases because of the high degree of homology in amino acid sequence, the structures are predicted to be similar to previously solved ones.

Amino acid sequence

These sequences were directly imported from either the SwissProt or the GenBank database. Amino acid sequences for all known species are included. The underline depicts the signal sequence removed to produce the mature protein.

Database accession numbers

In most cases GenBank, SwissProt, and PIR accession numbers are included. In the case of humans, MIM (Mendalian Inheritance in Man) database numbers were also included.

References

Only a selection of the most important references has been included. Most of these references represent critical observations in a given chemokine area.

CHEMOKINE RECEPTOR ENTRIES

The entries for chemokine receptors are categorized by their ligand binding properties. The α chemokine receptors are covered first, followed by the β chemokine receptors. Then follow entries for the receptors for the classical chemoattractants C5a and fMLP. DARC is listed under 'Miscellaneous' because if its promiscuous binding behavior. The receptor for platelet activating factor is also included in 'Miscellaneous'. Viral chemokine receptors include US28 and ECRF3. Finally, there are several orphan receptors of note listed.

The ligand binding and *in vitro* studies have been restricted to the human form of the receptor. Mutagenesis data is listed where the results have added to the understanding of the structure/function relationship of the receptor.

Figures of gene structures are drawn with open reading frames in black and untranslated regions in white; promoters are indicated by arrows and polyadenylation sites are depicted by an A over a vertical line. Protein structures are drawn in the classical seven-heptahelical form, with long rectangles representing the integral membrane regions, sequences above the rectangles representing the extracellular regions and below representing the cytoplasmic regions. Glycosylation sites are depicted by a ball on a stick. Putative phosphorylation sites are depicted by filled-in boxes. Disulfide bonds are indicated by the letters S connected by a dashed line.

Each chemokine receptor entry includes the following information where available.

Alternate names

Same as Chemokine entries

Family

Same as Chemokine entries

Homologs

This section describes homologs of a given receptor in different species.

Tissue Sources

Same as Chemokine entries. The expression patterns listed are those in which RNA or protein data were available.

In vitro biological effects

In vitro biological information are restricted to studies where the cloned receptor has been injected into *Xenopus* oocytes or transfected into a heterologous cell line for further study.

Animal models

This section describes any knockout or transgenic studies of the chemokine receptors.

Ligands and ligand binding studies

The ligand binding and *in vitro* biological information are restricted to studies where the cloned receptor has been injected into *Xenopus* oocytes or transfected into a heterologous cell line for further study. Other studies characterizing the functioning of a 'binding activity' have not been included since the effects of ligand-stimulation cannot be assumed to have been caused by the receptor.

Expression pattern

The expression patterns listed are those in which RNA or protein data were available.

Regulation of expression

This section describes the agents that cause changes in the expression at either gene or protein level.

Gene structure

Same as Chemokine entries

Protein structure

This section gives a general structure of the transmembrane protein, any known glyco-sylation sites, and potential serine/threonine phosphorylation sites.

In vitro signal transduction

This section highlights the intracellular events occurring after ligand binding.

Amino acid sequence

Same as Chemokine entries, except that the underline depicts transmembrane domain, where known.

Database accession numbers

Accession numbers are given for the most commonly used databases including GenBank, European Molecular Biology Laboratories (EMBL), SwissProt, Protein Identification Resource (PIR), and the National Center for Biotechnology (NCBI) gene identification number, along with the appropriate references, for the human receptor as well as any homologs whose sequence information is available.

3 Chemokines: an overview

HISTORICAL REVIEW

In number, degree of relatedness, and complexity, the chemokines are quite unlike anything that has previously been seen in the study of cytokines[1]. Unlike cytokines, which were first discovered by observation of soluble bioactivities, the majority of chemokines were identified by molecular cloning efforts. Molecular cloning of novel genes that encode small, secreted, and structurally related proteins continues to be the engine that is driving the expansion of the chemokine family. For almost three decades after the identification of the first chemokine (PF4 in 1961[2]) no other new chemokines were discovered. It was not until several new members were discovered that the common structural feature of the chemokine family, i.e., the presence of four conserved cysteines, was recognized. Since then the progress in chemokine research has been rapid and has led to the establishment of the chemokine superfamily.

GENERAL FEATURES OF CHEMOKINES

Chemokines are broadly divided into three families, α (CXC), β (CC), and γ (C), based on the presence and position of the conserved cysteine residues. In the members of the α family, the first two cysteines are separated by another amino acid, while in those of the β family they are placed next to each other[3]. Only two members of the γ family have been identified so far, and both of them contain one instead of two cysteines in their N-terminus. In the α and β families, two disulfide bonds are present between the first and third, and the second and fourth cysteine residues, respectively.

Understanding the cellular specificity of a given chemokine generally dominates the early phase of the research following its discovery. Experiments such as chemotaxis, calcium mobilization, and receptor binding are commonly used as measures of cell specificity. Such studies have led to a general consensus that the members of the α family are chemotactic to polymorphonuclear leukocytes (PMN) including neutrophils, eosinophils, and basophils. Members of the β family are generally chemotactic to mononuclear cells such as monocytes and lymphocytes[4]. Both members of the γ family are chemotactic to lymphocytes. Notable exceptions to this paradigm are the α-chemokines, IL-8, IP-10, and MGSA, which are reportedly chemotactic to T-lymphocytes and β chemokines MCP-1, MCP-3, RANTES, and MIP-1α which are also chemotactic to basophils[5].

The cellular specificity of a given chemokine depends on the presence or absence of specific receptors on the cell surface. Factors such as receptor sharing by several chemokines and a wide range in the number of surface receptors per cell has led to discrepancy in the ability of certain chemokines to induce biological effects and the observation of measurable binding of the chemokine in question. For example, studies addressing the receptor–ligand interactions of MCP-1 on different cell types failed to observe any measurable binding on T-lymphocytes despite reports of distinct biological effects[6]. It appears that very few receptors on the cell surface are sufficient to induce a potent biological effect such as chemotaxis. In most cases the presence of receptor mRNA and immunodetectable receptor protein in the membrane fraction are considered to be acceptable evidence for a cell type to be a specific target for a given chemokine.

Described below are some of the general biological effects that are common to both α- and β-chemokines. In addition, both families also possess some unique effects that are described under individual sections.

MIGRATION

Cell migration from circulation into tissue requires a coordinated sequence of events that involve

- morphological alterations in the responding cells which allow them to adhere to the endothelial lining of the blood vessels and transmigrate through the vessel wall
- dynamic remodeling of the cytoskeleton of the migrating cells
- expression of adhesion molecules that reversibly interact with extracellular matrix.

In addition, such migration requires cells to sense the changes in the concentration of the signal in the microenvironment to achieve directionality of movement. Chemokines appear to fit into this general paradigm in more than one way. It has been shown that several members of the chemokine family may be immobilized on the luminal surface of endothelium through cell surface proteoglycans[7]. Such immobilization allows them to be presented to the marginating leukocytes (*Figure 1*). Following interaction with the specific receptors on the leukocytes, chemokines induce remodeling of cytoskeleton such as reorganization of F-actin filaments that allows the cells to flatten and attain cellular polarization[8]. Changes in cell shape following chemokine stimulation of leukocytes also involve the formation of a

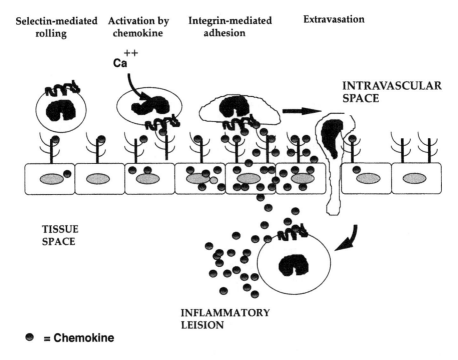

Figure 1.

cytoplasmic projection termed a 'uropod'[9]. The uropod is not simply a morphological structure but, rather, is likely to represent a specialized structure with important functions for motility and adhesion. Uropod formation is thought to be a critical event for transmigration of leukocytes. Once outside the circulation, leukocytes again utilize chemokines as signals that guide them to the tissues they need to reach, such as inflammatory lesions.

The *in vitro* assay used to measure the chemotaxis of leukocytes utilizes a Boyden chamber[10] or modified chambers that allow the researcher to utilize the popular 96-well format for high-throughput experiments. The principle of the assay is that the leukocytes are placed in a chamber above a polycarbonate filter membrane. The chemokine is placed in a chamber below the membrane and the cells are allowed to migrate through the membrane in response to the chemokine. Several published reports that clearly outline the procedural details are available[11-14].

CALCIUM MOBILIZATION

Interaction of chemokines with seven-transmembrane G-protein coupled receptors leads to rapid accumulation of intracellular free calcium in the responding cells[15]. Within 2 minutes the intracellular calcium returns to basal levels by redistribution into the calcium-sequestering compartments in the cells. Several studies address the sequence of events that lead to a rise in the cytosolic free calcium[16,17]. It appears that activation of phospholipase C (PLC) and conversion of phosphotidylinositol diphosphate (PIP_2) into diacylglycerol and inositol triphosphate (IP_3) is a critical step for the release of free calcium from the intracellular calcium stores. Release of intracellular stores is followed by entry of calcium from extracellular compartments into the cytosol. Calcium mobilization is a critical intracellular event for functional responses of chemokines such as chemotaxis, respiratory burst, and upregulation of adhesive interactions of leukocytes[18].

Calcium mobilization response is typically measured by labeling the leukocytes with calcium-sensitive dyes such as INDO-1, FURA-2, or FLUO-3 (Molecular Probes, Eugene, OR). Such dyes remain nonfluorescent unless they bind free calcium ions. Upon reversible binding to calcium, they develop fluorescence which can be measured by a fluorometer equipped with realtime continuous recording accessories. Several published reports that clearly outline the procedural details are available[19,20].

ADHESION

An important event in the trafficking of leukocytes to inflammatory sites is the adhesion of leukocytes to endothelial cells. There has been extensive characterization of cell surface adhesion molecules that mediate the adhesive interactions of different leukocytes with endothelial cells[21-23]. Chemokines regulate the expression of adhesion molecules on neutrophils, monocytes, lymphocytes and eosinophils. IL-8 causes shedding of L-selectin and upregulation of MAC-1 (CD11b/CD18) on neutrophils. MIP-1β induces CD8$^+$ T-cell adhesion to endothelium[24]. MIP-1α, MCP-1, and RANTES cause adhesion of monocytes to endothelium by upregulating both MAC-1 and p150,95 expression on monocytes[18]. Such expression results from membrane translocation of integrins from preformed intracellular stores. RANTES induces adhesion of eosinophilic cells to cultured TNF-α stimulated human umbilical vein endothelial cells (HUVECs)[25].

Chemokine effects on leukocyte adhesion are studied in two ways. First, by measuring the surface expression of adhesion molecules known to mediate leukocyte adhesion such as β_2 integrins (LFA-1, MAC-1, and p150,95). This can be done by fluorescent labeling of the integrins using FITC- or rhodamine-labeled monoclonal antibodies directed against the adhesion molecule and subsequent cell-associated fluorescence measurement using flow cytometry[18]. Alternatively, heterotypic adhesion of leukocytes to HUVECs can be measured using standard adhesion assays[26].

STRUCTURAL FEATURES OF α- AND β-CHEMOKINES

To date, three-dimensional structures of three α-chemokines, IL-8, GRO-α and PF4 and three β-chemokines, MIP-1β, RANTES, and MCP-1 have been solved by either multidimensional NMR or X-ray crystallography. The structures of all the monomers are very similar, as expected from the significant degree of sequence identity of these proteins (Figure 2)[27] . There are four significant differences between IL-8 and MIP-1β at monomer level. First, the conformation of the first disulfide bridge is a right-handed hook in IL-8 as opposed to a left-handed spiral in MIP-1β[28]. This is associated with the insertion in IL-8 of a residue between the first two cysteines and of two residues in the turn connecting β-strands 1 and 2. In contrast, the second disulfide bridge is almost perfectly superimposable between the two structures. Second, the helix extends for five residues further at the C-terminus in IL-8 compared to hMIP-1β. Third, the conformation of the turn connecting strands β_2 and β_3 differs around residues 46 and 47. Finally, the direction of the N-terminal residues preceding the first cysteine is completely different. The last three

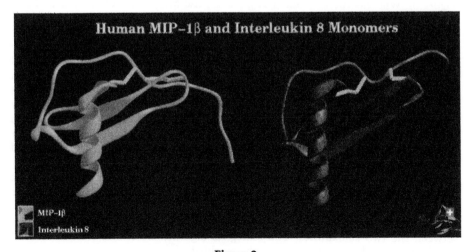

Figure 2.

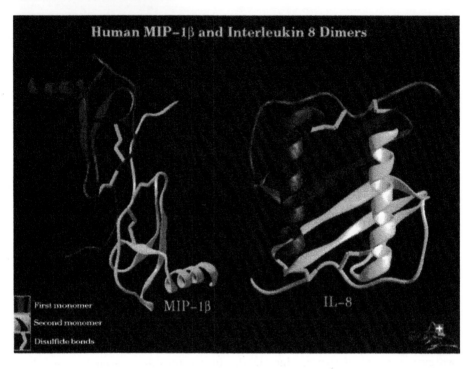

Human MIP–1β and Interleukin 8 Dimers

First monomer
Second monomer
Disulfide bonds

MIP–1β   IL–8

Figure 3.

differences at the monomer level are related to the different quaternary structures of the two proteins.

The quaternary structures of α- and β-chemokines, however, are entirely distinct and the dimer interface is formed by a completely different set of residues (*Figure 3*). Whereas the IL-8 dimer is globular, the MIP-1β dimer is elongated and cylindrical. PF4 is a tetramer comprising a dimer of IL-8 type dimers. The IL-8 dimer comprises a six-stranded antiparallel β-sheet, three strands contributed by each subunit, on top of which lie two anti-parallel helices separated by approximately 14 Å, and the symmetry axis is located between residues 26 and 26′ (equivalent to residue 29 of hMIP-1β) at the center of strands β_1 and β'_1. In contrast, in the hMIP-1β dimer the symmetry axis is located between residues 10 and 10′ which are part of an additional mini-antiparallel β-sheet formed by strands β_0 and β'_0; the two helices are 46 Å apart on opposite sides of the molecule; and strands β_1 and β'_1 are about 30 Å apart and located on the exterior of the protein. Calculation of the solvation free energies of dimerization and analysis of hydrophobic clusters strongly suggests that the formation and stabilization of the two different types of dimers arise from the burial of hydrophobic residues, and that the distinct quaternary structures are preserved throughout the two subfamilies. The differences in the quaternary structures between α- and β-chemokines provide an elegant explanation for the lack of receptor cross-binding between the two subfamilies. The biological significance of the dimeric form of the chemokines, however, has yet to be established[29].

α-CHEMOKINES

Although PF4 was the first discovered α-chemokine, it was not until the discovery of IL-8 that the importance of α-chemokines was recognized. Several α-chemokines with biological activities similar to those of IL-8 were discovered in rapid succession. These are NAP-2, GRO-α (MGSA), GRO-β (MIP-2α), GRO-γ (MIP-2β), ENA-78, and GCP-2. IL-8 shares sequence homology with these chemokines ranging from 24–46%.

Different members of the α family possess different biological activities. IL-8 is a potent chemoattractant for neutrophils. PF4 neutralizes the anticoagulant effect of heparin because it binds more strongly to heparin than to the chondroitin-4-sulfate chains of the carrier molecule. PF4 also appears to be chemotactic to fibroblasts in addition to neutrophils. MGSA was discovered as an autocrine growth factor for melanocytes. MGSA is also chemotactic to neutrophils. Platelet basic protein (PBP), which is the precursor of two active peptides – connective-tissue activating peptide III (CTAP-III) (also known as low-affinity platelet factor IV, LA-PF4) and β-thrombo-globulin (β-TG). CTAP-III is a growth factor that stimulates a variety of specific metabolic and cellular activities. EMF-1 is expressed constitutively in Rous sarcoma virus (RSV) infected cells and acts as an autocrine factor that promotes the growth of fibroblasts. Several members of the α family such as PF4 and platelet basic proteins β-TG, CTAP-III, and NAP-2 are found in the α granules of platelets and hence thought to provide a link between thrombosis, inflammation and wound healing – processes which are known to be physiologically coordinated[30].

α-Chemokines and angiogenesis

α-Chemokines display disparate angiogenic activity depending upon the presence or absence of the ELR motif. Those containing the ELR motif, such as IL-8 and MGSA, are found to be potent angiogenic factors, inducing both *in vitro* endothelial chemotaxis and *in vivo* corneal neovascularization[31]. In contrast, the α-chemokines lacking the ELR motif, PF4, IP-10, and MIG, not only fail to induce significant *in vitro* endothelial cell chemotaxis or *in vivo* corneal neovascularization but are also potent angiostatic factors in the presence of either ELR-α-chemokines or the unrelated angiogenic factor, basic FGF[32]. Additionally, mutant IL-8 proteins lacking the ELR motif demonstrate potent angiostatic effects in the presence of either ELR-α-chemokines or basic FGF. In contrast, a mutant of MIG engineered to contain the ELR motif can induce *in vivo* angiogenic activity. These findings suggest a functional role of the ELR motif in determining the angiogenic or angiostatic potential of α-chemokines, supporting the hypothesis that the net biological balance between angiogenic and angiostatic α-chemokines may play an important role in regulating overall angiogenesis.

β-CHEMOKINES

The first human β-chemokine was identified by differential hybridization cloning and was named LD78[33]. Several cDNA isoforms of a closely related human chemokine, Act-2, were later described, and two similar proteins, MIP-1α and MIP-1β, were purified from culture medium of LPS-stimulated mouse macrophages. On the basis of more than 70% amino acid identity, the murine and human proteins are considered homologs, and the terms human MIP-1α and MIP-1β are commonly used instead of LD78 and Act-2[34].

MCP-1 is perhaps the most studied β-chemokine and its history can be traced to the discovery of murine *JE*, an immediate early gene, in 1983[35], and its subsequent cloning and characterization[36]. MCP-1 was discovered by molecular cloning as a transcript induced rapidly in fibroblasts by PDGF. Because of the potent and specific chemotacitic effect of MCP-1 on monocytes, much emphasis was placed on understanding the biology of MCP-1. Identification of specific receptors for MCP-1 on monocytes further strengthened the hypothesis that MCP-1 could play a significant role in the tissue infiltration of monocytes as seen in many chronic inflammatory lesions. Purification and sequencing of novel monocyte chemotactic activities from the human osteosarcoma cell line MG63 led to identification of two related proteins, MCP-2 and MCP-3 with 62% and 73% amino acid identity to MCP-1, respectively[11]. MCP-3 was later identified to share the receptors with MCP-1 and acts as a very potent histamine secretogogue of human basophils[37]. The biological role of MCP-2 still appears to be in question.

RANTES is another very extensively studied β-chemokine. RANTES was also originally identified by molecular cloning as a transcript expressed in T cells but not in B cells[38]. RANTES generated a lot of interest among the chemotactic research community primarily because of its potent chemotactic effect on lymphocytes. It is thought to be a parallel of MCP-1 and together they are thought to be responsible for monocyte and lymphocyte influx into inflammatory lesions.

β-Chemokines as HIV suppressive factors

Recent evidence suggests that the chemokines RANTES, MIP-1α, and MIP-1β are major HIV-suppressive factors (HIV-SF) produced by CD^{8+} T cells[39]. Two active proteins purified from the culture supernatant of an immortalized CD^{8+} T-cell clone revealed sequence identity with human RANTES and MIP-1α. RANTES, MIP-1α, and MIP-1β are released by both immortalized and primary CD^{8+} T cells. HIV-SF activity produced by these cells was completely blocked by a combination of neutralizing antibodies against RANTES, MIP-1α, and MIP-1β. Human RANTES, MIP-1α, and MIP-1β induced a dose-dependent inhibition of different strains of HIV-1, HIV-2, and simian immunodeficiency virus (SIV). These data may have relevance for the prevention and therapy of AIDS.

References

[1] Schall, T. (1994) The chemokines. In *The cytokine handbook* (ed. A. Thomson). Academic Press, London.

[2] Deutsch, E. and Kain, W. (1961). In *Blood platelets* (ed. S.A. Johnson), Little Brown, Boston, MA.

[3] Yoshimura, T. and Leonard, E.J. (1990) *J. Immunol.* 145, 292–297.

[4] Oppenheim, J.J. et al. (1991) *Annu. Rev. Immunol.* 9, 617–648.

[5] Bacon, K. and Schall, T. (1996) *Int. Arch. Allergy Immunol.* 109, 97–109

[6] Carr, M.W. et al. (1994) *Proc. Natl. Acad. Sci. USA* 91, 3652–3656.

[7] Tanaka, Y. et al. (1993) *Nature* 361, 79–82.

[8] del Pozo, M. et al. (1996) *J. Cell Biol.* 121, 495–508.

[9] Vaddi, K. and Newton, R.C. (1993) *J. Leuk. Biol.* 55, 756–762.

10 Falk,W. et al. (1980) *J. Immunol. Methods* 33, 239–245.
11 Junger, W.G. et al. (1993) *J. Immunol. Methods* 160, 73-79
12 Bacon, K. et al. (1988) *Br. J. Pharmacol.* 95, 966–980.
13 Harvath, L. et al. (1980) *J. Immunol. Methods* 37, 39–45.
14 Samantha, A. et al. (1990) *J. Biol. Chem.* 265, 183–189.
15 Murphy, P.M. (1994) *Annu. Rev. Immunol.* 12, 593–633.
16 Sozzani, S. et al. (1993) *J. Immunol.* 147, 2215–2221.
17 Bacon, K. et al. (1995) *J. Immunol.* 154, 3654–3666.
18 Vaddi, K and Newton, R.C. (1994) *J. Immunol.* 153, 4721–4732.
19 Naccache, P.H. et al. (1989) *J. Immunol.* 142, 2438–2444.
20 Seifert, R. et al. (1992) *Mol. Pharmacol.* 42, 227–236.
21 McKay C.R. and Imhof, B.A. (1993) *Immunol. Today* 14, 99–101
22 McEver, R.P. (1992) *Curr. Opinion. Biol.* 4, 840–860
23 Hynes, R.O. (1992) *Cell* 65, 859–870
24 Gilat, D. et al. (1994) *J. Immunol.* 153, 4899–4906
25 Ebisawa, M. (1994) *J. Immunol.* 153, 2153–2160
26 Bevilacqua, M. et al. (1987) *Proc. Natl. Acad. Sci. USA* 84, 9238–9245.
27 Baggiolini, M. et al. (1994) *Adv. Immunol.* 55, 97–179
28 Lodi, P.J. et al. (1994) *Science* 263, 1762–1767.
29 Clore, G.M. and Gronenborn, A.M. (1995) *FASEB J.* 9, 57–62.
30 Miller, M.D. and Krangel, M.S. (1992) *CRC Crit. Rev. Immunol.* 12, 17–46.
31 Koch, A.E. et al. (1992) *Science* 258, 1798.
32 Strieter, R.M. et al. (1995) *Biochem. Biophys. Res. Comm* 210, 51–57.
33 Nakao, M. et al. (1990) *Mol. Cel. Biol.* 10, 3646–3658.
34 Baggiolini, M. and Dahinden, C.A. (1994) *Immunol. Today* 15, 127–133.
35 Cochran, B.H. et al. (1983) *Cell* 33, 939–947.
36 Rollins, B.J. (1991) *Cancer Cells* 3, 517–524.
37 Dahinden, C.A. et al. (1994) *J. Exp. Med.* 179, 751–756.
38 Schall, T. (1991) *Cytokine* 3, 165–183.
39 Cocchi, F. et al. (1995) *Science* 270, 1811.

4 Chemokine receptors: an overview

HISTORICAL REVIEW

Chemoattractant receptors were first discovered through their role in bacterial toxicity. Bacterially produced formyl peptides were found to be potent activators of immune cells, leading to the characterization and eventual cloning of formyl peptide receptors, including the most characterized member of the group, the fMLP receptor. Among the chemokine receptors, the IL-8 receptors were the first to be studied at the molecular level, resulting in a large body of mutagenesis data relating structure to function, especially ligand-binding specificity.

GENERAL STRUCTURAL FEATURES

Chemokine receptors are integral membrane glycoproteins (via *N*-linked glyco-sylation) with heptahelical structure composed of seven helices snaking through the membrane with the N-terminus on the extracellular face and the C-terminus on the cytoplasmic side of the membrane. In general, these receptors average at 350 amino acids with a calculated molecular weight of ~40 kD.

RECEPTOR–LIGAND INTERACTIONS

Siciliano et al. propose a two-site binding model[1] of C5a bound to the C5a receptor, and this model is probably relevant for at least some of the other family members as well. Mutagenesis studies show that the truncated C5a receptor did not bind C5a but did bind activating peptides derived from C5a. Site 1 is at the N-terminus of the receptor and is proposed to undergo a conformational change that exposes site 2 which interacts with the C-terminus of the ligand.

The ternary complex model[2] of receptor, ligand, and G-protein describes the interaction of the signal transduction machinery with the receptor and the effect of receptor occupancy on the complex. There are high and low affinity states of the receptor, thought to be due to association of the receptor with G protein in the presence (high affinity) and absence (low affinity) of ligand.

CHEMOKINE RECEPTORS AS G-PROTEIN COUPLED RECEPTORS

There are more than 100 G-protein coupled receptors, with ligands ranging from small molecules to moderate sized proteins. The chemokine receptors are smaller members (about 350 amino acids) with more limited tissue distribution (immune cells and associated tissues).

These receptors are coupled to regulatory GTP binding proteins, thus leading to signal transduction and a series of early events (calcium influx, PI turnover) and late events (chemotaxis, actin polymerization). Signalling of these receptors has been reviewed recently and will not be covered in detail here[3].

Mutagenesis studies suggest that the area of interaction with the G protein can be localized to the third cytoplasmic loop and the C-terminal tail. Some GPCRs have been shown to be phosphorylated by GPCR-specific serine/threonine kinases. N-linked glycosylation sites are identified by the primary sequence of the receptors.

Chemokine receptors have several features unique to their members. Most have an acidic N-terminal tail containing a Pro-Cys motif. There is a conserved Cys in the third extracellular domain (EC3) that is implicated in disulfide bonding. The third cytoplasmic loop is shorter than in other branches of the GPCR family.

Rhodopsin is the prototypical GPCR. The GPCR superfamily has grown to include six families. These are:

- rhodopsin and adrenergic-like receptors
- calcitonin-like
- metabotropic
- pheromone-like
- cAMP-like
- other

The chemokine receptors are categorized in the first family and themselves can be further subdivided into leukocyte chemokine receptors that bind α- or β-chemokines but not both; viral homologs of leukocyte chemokine receptors and Duffy erythrocyte antigen which binds both α- and β-chemokines.

EXPERIMENTAL STUDIES

Many studies involve examining the receptor in a hematopoietic cell line or transfecting a cDNA-containing expression vector into a heterologous cell line.

CURRENT STATUS

Much of the structural information known about GPCRs is based on detailed studies of bacteriorhodopsin and is reviewed in Ref. 4. A good resource for G-protein coupled receptors is Gert Vriend's World Wide Web site (http://swift.embl-heidelberg.de/7tm/htmls/GPCR.html). It has the currently available DNA and protein information for each member of the GPCR family, organized into their subgroups, as well as three-dimensional models of the receptors, where available.

The role of the chemokine receptors in HIV infection has recently been described for LESTR (fusin)[5]. The latest addition to the beta chemokine receptors, CC-CKR-5[6], receptor for MIP-1α, MIP-1β, and RANTES, has recently been implicated playing a role in HIV infection as well[7].

ORPHAN 'CHEMOKINE' RECEPTORS

Several cDNAs have joined the chemokine receptor family through homology to the existing members although as yet they have no known ligand or function.

VIRAL CHEMOKINE RECEPTORS

Nature has utilized the chemokine receptor in virus biology as well as in host immunity. Several proteins with homology to chemokine receptors have been shown in transfected cells to bind known chemokines, even functioning to transmit a signal. This has led some to suggest that the virus that produces such a protein in an infected cell is using cellular mechanisms already in place to do its bidding. This is especially evident in the herpes virus family, in which US28 and ECRF3 as well as several recently cloned receptor homologs will further our understanding of viral infiltration as well as chemokine receptor function.

General reviews

Ahuja, S.K. et al. (1994) *Immunol. Today* 15, 281–287.
Donnelly, D. and Findlay, J.B.C. (1994) Current Opinion in Structural Biology 4, 582–589.
Gerard, C. and Gerard, N.P. (1994) *Annu. Rev. Immunol.* 12, 775–808.
Horuk, R. (1994) *Trends Pharmacol. Sci.* 15, 159–165.
Horuk, R (1994) *Immunol. Today* 15, 169–174.
Kelvin, D.J. et al. (1993) *J. Leukoc. Biol.* 54, 604–612.
Kelvin, D.J. et al. (1993) *Adv. Exp. Med. Biol.*. 351, 147–153.
Murphy, P.M. (1994) *Annu. Rev. Immunol.* 12, 593–633.
Murphy, P.M. et al. (1994) *Infect. Agents Dis.* 3, 137–154.
Schall, T.J. et al. (1993) *Adv. Exp. Med. Biol.* 351, 29–37.

References

[1] Siciliano, S.J. et al. (1994) *Proc. Natl. Acad. USA* 9, 1214–1218.
[2] De Lean, A. et al. (1990) *J. Biol. Chem.* 255, 7108–7117.
[3] Brass, L.E. et al. (1992) *J. Biol. Chem.* 267, 13795–13798.
[4] Strader, C.D. et al. (1994) *Annu. Rev. Biochem.* 63, 101–132.
[5] Feng, Y. et al. (1996) *Science* 272, 872–877.
[6] Samson, M. et al. (1996) *Biochem.* 35, 3362–3367.
[7] Doranz, B.J. et al. (1996) *Cell* 85, 1149–1158.

THE CHEMOKINES

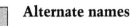

Alternate names

Neutrophil activating protein-1[1], monocyte-derived neutrophil chemotactic factor (MDNCF), LUCT (lung giant cell carcinoma-derived chemotactic protein), AMCF-I (alveolar macrophage-derived chemotactic factor (pig), TPAR1 (TPA repressed gene 1), tumor necrosis factor stimulated gene 1 (TSG1)

Family

α family (CXC family)

Molecule

IL-8 is the prototypic and most extensively studied chemokine. Because of its potent neutrophil chemotactic effect, IL-8 is thought to be the primary regulatory molecule of acute inflammatory states. Recently, using high-affinity neutralizing anti-IL-8 antibodies, the regulatory effect of IL-8 on tissue infiltration of neutrophils has been identified. Another important effect of IL-8 is promotion of angiogenesis. Structural studies on IL-8 led to the identification of the ELR motif that is critical to the chemotactic effect of several members of the α-chemokine family.

Tissue sources

Monocytes, macrophages[2], CD 4$^+$, CD 8$^+$ and CD45RA$^+$ lymphocytes, epithelial cells, epidermal keratinocytes, human $\gamma\delta$ T cell lines, human renal cell carcinoma cell line (TC-2), human melanoma cells, human malignant fibrous histiocytoma cell line (MMF-1), human glioblastoma cell lines, human peritoneal mesothelial cells, cultured chorion and decidual cells, gingival fibroblasts, human mesangial cells, cytokeratin-positive renal cortical epithelial cells.

Target cells

Neutrophils, T-lymphocytes[3], basophils, eosinophils, keratinocytes, HUVECs.

Physicochemical properties

Property	Human	Rabbit	Sheep	Pig	Dog
pI (mature)	8.6	~8	~8	~8	~8
Signal	1–22	1–22	1–22	1–22	1–22
Aminoacids					
Precursor	99	101	101	103	101
Mature	77 (endothelial)				
	72 (monocyte)	78	78	80	78
Disulfide bonds a.a.	34–61	34–61	34–61	34–61	34–61
a.a.	36–77	36–77	36–77	36–77	36–77
Glycosylation sites	0	0	0	0	0
Molecular weight (predicted)	11 098	11 402	11 292	11 633	11 280

23

Transcription factors

Several members of the NF-κB family, including p65, p50, p52, and c-Rel, can bind to the IL-8 promoter. Strong cooperative binding of C/EBP to its site in IL-8 promoter is seen only when NF-κB is bound to its adjacent binding site. A positive synergistic activation through the C/EBP binding site and inhibition through the NF-κB binding site occur by combinations of C/EBP and NF-κB. The ultimate regulation of IL-8 gene expression depends on the ratio of cellular C/EBP and NF-κB[4].

RelA is a major component in the κB binding complexes to IL-8-κB site which binds RelA and c-Rel and NF-κB2 homodimers, but not to NF-κB1 homodimers or heterodimers of NF-κB1-RelA.

cRel-p65 heterodimers from both monocytic and endothelial cells bind to the κB-like site in the IL-8 gene with a consensus sequence, 5'-HGGARNYYCC-3'[5].

Transcription of the IL-8 gene requires the activation of either the combination of NF-κB and AP-1 or that of NF-κB and NF-IL6, depending on the type of cells. The region from -98 to $+44$ of the IL-8 gene is sufficient to confer both inducibility of IL-8 by TNF and inhibition by simultaneous treatment with IFN-β. Inhibition of TNF- or IL-1-induced IL-8 gene expression by IFN-β or IFN-α is also observed when a DNA fragment containing only the NF-IL-6 and NF-κB sites (positions -94 to -70) is placed upstream of the homologous or a heterologous minimal promoter. The NF-κB element alone is sufficient to confer inhibition of TNF-induced IL-8 gene transcription by IFN-β[6].

Both NF-κB and C/EBP-like *cis*-elements located at -94 to -71 bp of the *IL-8* gene are essential and sufficient for the induction of the *IL-8* gene by protein X of hepatitis B virus and by cytokines such as IL-1 and TNF-α.

Regulation of expression

Stimulus	Cell type	Response
Stress		
Hypoxia	Endothelial cells	IL-8 release
		IL-8 mRNA accumulation
		NF-κB activation
Anoxic preconditioning and oxygen stress	Monocytes	Augmented IL-8 protein and gene expression
Ultraviolet B radiation	Keratinocyte cell line, A431	Up-regulated expression of IL-8
Viruses		
Respiratory syncytial virus	Bronchial epithelial cells	IL-8 mRNA expression
Influenza A virus	Rat kidney NRK-52E cells	IL-8 mRNA expression
	Human airway epithelial cells	IL-8 mRNA expression
HIV-1	LPS-stimulated monocytes	Elevation of induction of IL-8 expression
	IFN-γ treated monocytes	Elevation of IL-8 expression

Stimulus	Cell type	Response
Protein X of the hepatitis B virus (HBV-X)	HBV-X-transfected cells	Enhanced IL-8 mRNA expression and IL-8 production

Bacteria

Stimulus	Cell type	Response
Staphylococcus aureus Capsular polysaccharide types 5 and 8 (CP5 and CP8)	Human epithelial κB cells, endothelial cells monocytes	IL-8 production
Mycobacterium tuberculosis and its cell wall components lipoarabinomannan (LAM), lipomannan, and phosphoinositolmannoside	Alveolar macrophages	Stimulation of IL-8 protein release and mRNA expression
Mycobacterium tuberculosis	Human monocyte cell lines	Secretion of IL-8 observed following phagocytosis
Viable staphylococci	Mesothelial cells	IL-8 production
Helicobacter pylori[a] (NCTC 11637)	Gastric epithelial cell lines (KATO-3, ST42, AGS)	Induction of IL-8 secretion
	Human epithelial cell line Hep-1	Induction of IL-8 expression
Pseudomonas aeruginosa (piliated wild-type organisms purified pili, or antibody to the pilin receptor, asialoGM1)	Immortalized airway epithelial cells	Evokes IL-8 production
Propionibacterium acnes	Monocytic cell lines and peripheral blood mononuclear cells	IL-8 induction
LPS	Human peripheral whole blood	IL-8 production peaking at 6–12 h
	Nasal but not bronchial or lung tissue-derived fibroblasts	Induction of IL-8 expression

Others

Stimulus	Cell type	Response
All-*trans* retinoic acid	Human epithelial ovarian cancer cell line HOC-7	IL-8 mRNA expression
1,25(OH)$_2$-vitamin D$_3$	Keratinocytes, fibroblasts and mononuclear cells but not endothelial cells	Inhibition of IL-1-α induced IL-8 production and mRNA expression
ETH615, an inhibitor of leukotriene biosynthesis	Human peripheral blood mononuclear cells	Inhibition of LPS-induced expression of IL-8 mRNA and production
ADP or epinephrine-activated platelets	Endothelial cells	IL-8 secretion mediated by IL-1
Staurosporine	IFN-γ primed neutrophils	Induction of IL-8 mRNA
Cadmium	Human monocytes	Elevated expression of IL-8 protein and mRNA
Isoprenoid synthesis inhibitors such as lovastatin or compactin	Monocytic cell line (THP-1)	Suppression of IL-8 production induced by LPS, GM-CSF, and PMA
Histamine	Human endothelial cells	Increased secretion and mRNA expression of IL-8
Dexamethasone	Resting and IL-1 or TNF-α stimulated human airway epithelial cells	Suppression of IL-8 mRNA expression and protein synthesis

Stimulus	Cell type	Response
Dexamethasone	IL-1 stimulated human fibrosarcoma cell line, 8387	Suppression of IL-8 production
	Human mesenchymal cells	Suppression of IL-8 production
	Peripheral blood monocytes and alveolar macrophages	Suppression of IL-8 production
Nedocromil sodium	Human bronchial epithelial cells	Reduction of the IL-1-induced release of IL-8
Sodium aurothiomalate, D-penicillamine and sulphasalazine	Endothelial cells	Suppression of IL-8 production at low concentrations and enhancement at high concentrations
Hydrocortisone	Endothelial cells	Suppression of IL-8 production
5-Aminosalicylic acid	Endothelial cells but not monocytes	Induction of increased IL-8 release
Leukotriene B$_4$	Neutrophils	Stimulation of synthesis and secretion of biologically active IL-8
PGE$_2$	Peripheral blood monocytes but not alveolar macrophages	Suppression of LPS-induced IL-8 production
IL-1 receptor antagonist protein (IRAP)	LPS-stimulated human whole blood	Suppression of IL-8 production
NO donors, 3-morpholino-sydnonimine (SIN-1), S-nitroso-N-acetylpeni-cillamine (SNAP), and S-nitroso-L-glutathione (SNOG)	Human melanoma cell line, G361	IL-8 secretion and promoter activity
Sulfatides with sulfation of the galactose ring of the glycolipid	Human monocytes	Induction of IL-8 expression
Human stroma-free hemolysate, purified adult hemoglobin Ao and oxidized HbAo	Human mononuclear cells	IL-8 secretion
Calyculin A, a potent type 1/2A protein serine/threonine phosphatase inhibitor	Synovial cells	Induction of IL-8 mRNA expression
Auranofin and staurosporine, inhibitors of PKC	PMA-stimulated keratinocytes	Inhibition of IL-8 production and mRNA expression
	IL-1 and TNF-stimulated keratinocytes	Potentiation of the minimal IL-8 protein and mRNA
Calcium ionophore, A23187	Human monocytic cells	Induction of IL-8 gene expression and protein secretion
	Human peripheral blood eosinophils	Production and release of IL-8
N-ethylcarboxamido-adenosine (NECA)	Mast cell line HMC-1	IL-8 secretion
Theophylline and enprofylline	NECA-stimulated mast cell line HMC-1	Inhibition of IL-8 release

Stimulus	Cell type	Response
Monocytes	Unstimulated HUVECs	Increased production of IL-8
Neurotoxic fragment (25–35) β-Amyloid A β	Cultured monocytes	Stimulation of IL-8 mRNA and protein expression
Urinary trypsin inhibitor	LPS-treated HL60 cells	Down-regulation of IL-8 gene expression
Cholecystokinin	Macrophages	Induction of IL-8 production
Polyspecific IgG	Human monocytes	Stimulation of IL-8 production
Contact sensitizer (2,4-dinitrofluorobenzene), tolerogen (5-methyl-3-n-pentadecylcatechol) and irritant (sodium lauryl sulfate)	Human keratinocytes	Induction of IL-8 mRNA but not protein secretion
Adherence to plastic or extracellular matrix, such as collagen type I or fibronectin	Monocytes Alveolar macrophages	Stimulation of IL-8 secretion and gene expression
Calcitriol, 9-cis-retinoic acid, and sodium butyrate	PMA-stimulated monocytes	Potentiation of IL-8 mRNA expression
Cytokines		
IFN-γ	Resting neutrophils	Down-regulation of constitutive IL-8 mRNA levels
	TNF- or LPS-activated neutrophils	Down-regulation of elevated expression of IL-8 mRNA
	IL1-stimulated monocytes	Inhibition of IL-8 expression
	LIF-stimulated monocytes	IL-8 induction inhibited
IFN-β	TNF-stimulated monocytes	Enhancement of IL-8 gene expression
IFN-α	LPS- and TNF-stimulated neutrophils	Elevation of IL-8 expression inhibited
IL-10	Neutrophils	Inhibition of IL-8 synthesis and enhanced degradation of LPS-induced IL-8 mRNA
	Monocytes	Inhibition of LPS-induced transcription of IL-8 gene, enhanced degradation of LPS-induced IL-8 mRNA
IL-1β	Human peritoneal mesothelial cells	IL-8 production
	Human gingival fibroblasts	IL-8 production
IL-1β, TNF-α and GMCSF	Lung cancer cell lines, RERF-LC-MS, RERF-LC-OK, A549, and YO-88	Induction of IL-8 production
IL-1-β and TNF-α[b]	Human synovial fibroblasts	Induction of IL-8 production
	Human transitional cell carcinomas and renal cell carcinomas	Release of IL-8 protein
	Human corneal endothelial and stromal cells	
	Smooth muscle cells	

Stimulus	Cell type	Response
IL-1β and TNF-α[b]	Pulmonary fibroblasts	Induction of IL-8 production
	Glomerular mesangial cells Pulmonary type II-like epithelial cell line, A549 Neural-derived retinal pigment epithelial cells Alveolar macrophages	
Monocyte membrane anchored IL-1α	Human endothelial cells	Juxtacrine induction of IL-8 expression
TNF-α	Human peritoneal mesothelial cells	IL-8 production
	Bronchial epithelial cells	Up-regulation of IL-8 expression
Anti-TNF receptor (p55) agonistic antibody, Htr-9	Human vascular endothelial cells	Release of IL-8
GMCSF	Peripheral blood monocytes and neutrophils	Induction of IL-8 mRNA accumulation and protein secretion
IL-3	Peripheral blood monocytes	Induction of IL-8 mRNA accumulation and protein secretion
Recombinant CD40 ligand	Primary and cultured Hodgkin and Reed–Sternberg (H-RS) cells	Induction of IL-8 secretion
LIF	Macrophages Myeloma cells	Induction of the expression of IL-8 mRNA

[a] CagA/cytotoxin positive strains of *H. pylori* induce higher IL-8 secretion than CagA/cytotoxin negative strains in all the three positive gastric epithelial cell lines.
[b] Preincubation of the cells with cycloheximide superinduces the level of IL-8 mRNA stimulated by TNF-α and IL-1β.

Expression in disease

Disease	Tissue	Comment
Reperfusion injury of ischaemic lungs	Lung	Neutrophil infiltration and destruction of pulmonary structure
Chorioamnionitis	Amniotic fluids	IL-8 protein detected
	Cytotrophoblast, syncytiotrophoblast and Hofbauer cells of the placenta, decidual stromal cells, decidual lymphocytes and endometrial gland cells	IL-8 protein and mRNA detected
Human pregnancy	Unstimulated CD16–CD56 bright NK cells in decidua	Levels comparable to the IL-8 production by LPS-stimulated peripheral blood mononuclear cells
Human malignant astrocytoma	Peritumoral brain tissue	Not present in normal brain
β-Thalassemia	Serum	IL-8 protein detected
Adult respiratory distress syndrome (ARDS) and pneumonia	Bronchoalveolar lavage fluid	Elevated IL-8 production

Disease	Tissue	Comment
Chronic sinusitis and allergic rhinitis	Maxillary mucosa	IL-8 gene expression
Pediatric cardiopulmonary bypass	Heart and skeletal muscle	IL-8 gene expression
Cystic fibrosis	Bronchoalveolar lavage fluid	IL-8 protein expression
Psoriasis	Psoriatic lesional extracts	IL-8 protein expression
Ulcerative colitis	Colonic mucosa	Diffusely distributed
Crohn's disease[a]	Colonic mucosa	Focal distribution pattern: cells expressing IL-8 are mainly located at the base of ulcers, in inflammatory exudates on mucosal surfaces, in crypt abscesses, and at the border of fistulae
Whole blood hemolytic transfusion reactions following addition of ABO-incompatible red blood cells	Serum	IL-8 production related to the degree of hemolysis; inhibited by inactivation of complement
Idiopathic pulmonary fibrosis or sarcoidosis	Human alveolar macrophages from brochoalveolar lavage	Expression of IL-8 protein
Pyuria, *Escherichia coli* infection of the urinary tract	Urine	Expression of IL-8 protein
Chronic airway disease and *Pseudomonas aeruginosa* infection	Bronchoalveolar lavage fluid	Expression of IL-8 protein
Acute myeloblastic leukemia	Leukemic cells	Expression of IL-8 protein
Mild asthmatics	Bronchial epithelium	Down-regulated by inhaled dipropionate and beclomethasone hydrocortisone
Localized and septicemic *Pseudomonas pseudomallei* infection	Plasma	Expression of IL-8 protein
Clinical sepsis	Serum	Well correlated with lactate, IL-6, elastase-α_1-antitrypsin, and C3a levels and inversely with leukocyte and platelet numbers and mean arterial pressure
Pseudophakic bulbous keratopathy	Cornea	Expression of IL-8
Pulmonary tuberculosis	Macrophages and in lavage fluid	IL-8 mRNA is found in alveolar macrophages
B-cell chronic lymphocytic leukemia	Serum	Not found in B-cells from hairy cell leukemia
Human periodontal infections	Junctional epithelium adjacent to the infecting microorganisms	Correlates with neutrophil infiltration
Rheumatoid arthritis (RA)	Leukocytes	
Herpes simplex virus infection	Human corneal keratocytes	Not in human corneal epithelial cells

Disease	Tissue	Comment
Otitis media with effusion	Middle ear effusion	High levels of IL-8
Animal studies		
Endotoxin-induced pleurisy	Lung tissue	Rabbit model of pleurisy

[a] Analysis of semiserial sections points to macrophages, neutrophils, and epithelial cells as possible sources of this cytokine in active inflammatory bowel disease.

In vivo studies

Animal studies

In IL-8 transgenic mice, elevated serum IL-8 levels correlate with proportional increases in circulating neutrophils and proportional decreases in L-selectin expression on the surface of blood neutrophils. No change in the expression of the β_2-integrins MAC-1 and LFA-1 is apparent on peripheral blood neutrophils of the IL-8 transgenic mice. Additionally, L-selectin expression on bone marrow neutrophils and neutrophil precursors was normal in all transgenic lines. IL-8 transgenic mice demonstrate an accumulation of neutrophils in the microcirculation of the lung, liver and spleen. There is no evidence of neutrophil extravasation, plasma exudation or tissue damage in any IL-8 transgenic mice. Neutrophil migration into the inflamed peritoneal cavity is severely inhibited in IL-8 transgenic mice[7].

Application of IL-8 causes corneal neovascularization *in vivo* in rabbits.

In rats, the intracerebroventricular administration of IL-8 decreases short-term (2 h) food intake.

In vivo administration of IL-8 induces granulocytosis and the release of immature white blood cells into the circulation[8].

Superfusion of dog tracheal rings with culture supernatants from *Staphylococcus aureus* or *Pseudomonas aeruginosa* induces IL-8 production and subsequent neutrophil recruitment.

In a rabbit model of myocardial infarction, coronary artery occlusion followed by reperfusion for 4.5 h leads to delayed high-level expression of IL-8. Intravenous IL-8 in rabbits reduces granulocyte recruitment to inflammatory sites by inhibiting function(s) necessary for transmigration that are independent of L-selectin and subsequent to rolling.

Intraperitoneal administration of IL-8 in mice mobilizes GM-CFU within 15 min. Administration of mononuclear cells obtained from IL-8-treated mice results in 100% survival of lethally irradiated mice[8].

Intracellular signaling

IL-8-stimulation causes activation of GTPase by increasing V_{max} but not by decreasing K_m. High-affinity binding of GTP[^{35}S] to neutrophil plasma membranes is stimulated half-maximally and maximally (up to 5-fold) by IL-8 at about 10 nM and 100 nM respectively[9].

IL-8 stimulates the breakdown of 1-*O*-[3H]alkyl-2-acyl-*sn*-glycero-3-phospho-choline ([3H]EAPC) and the formation of 1-*O*-[3H]alkyl-2-acyl-phosphatidic acid ([3H]-EAPA) in human neutrophils. In addition IL-8 also activates phospholipase D and stimulates the metabolism of choline-containing phosphoglycerides in human neutrophils in a manner that is sensitive to PT and ethanol.

IL-8 but not other α-chemokines enhances the formation of phosphatidyl-ethanol in a manner inhibitable by PT and the tyrosine kinase inhibitors erbstatin and herbimycin A.

A latent outwardly rectifying K$^+$ conductance, GkOR, is elicited within seconds by IL-8 in the murine J774 monocytic cell line.

PKC and calmodulin-like protein dependent phosphorylation of a cytosolic protein, p64, is induced by IL-8 stimulation of neutrophils.

IL-8 causes ligand-dependent activation of endogenous PLC with the involvement of G-protein α subunits Gα 14, Gα 15, or Gα 16, but not Gα 9 or Gα 11. IL-8 receptors interact with PT-sensitive G proteins or with the recombinant G protein Gi to release free $βγ$ subunits that specifically activate the $β_2$ isoform of PLC.

IL-8 activates extracellular signal regulated kinase (ERK1) in human neutrophils.

IL-8 induces phosphorylation of IL-8RA overexpressed in a rat basophilic leukemia cell line, RBL-2H3. Phosphorylation of IL-8RA correlates with its desensitization as measured by GTPase activation and calcium mobilization[10].

IL-8 stimulates polyphosphoinositide hydrolysis and generation of IP$_3$ in human peripheral blood lymphocytes which can be partially inhibited by the tyrosine kinase inhibitor, genistein.

Receptor binding characteristics

Human neutrophils

^{125}I-IL-8 binds at 0–4°C to neutrophils at 64 500 receptors/cell with an apparent K_d of 0.18 nM. Unlabeled IL-8, NAP-2, and MGSA/GRO compete with ^{125}I-NAP-1/IL-8 for binding to human neutrophils[11].

Neutrophil membranes

^{125}I-IL-8 binds to isolated neutrophils membranes in a rapid and reversible manner with an equilibrium dissociation constant (K_d) of 5.0–12.4 nM and number of receptors of 1.58–5.90. 10^{10} receptors/mg protein. Treatment of membranes with the nonhydrolyzable analogs of GTP, GMP-PNP and GTPγS, inhibits the binding of IL-8 indicating conversion of the receptors from a high affinity state to a low affinity state in the presence of guanine nucleotides.

Basophils

IL-8 binds to specific receptors on basophils at 3500–9600 receptors with a mean K_d value of 0.15 nM[12].

HL-60 cells

Steady state binding of IL-8 with a K_d value of 0.5 nM and receptor number of 7400 for differentiated HL-60 cells is reported.

Human embryonic leukemia (HEL) cells

Saturation binding studies with ^{125}I-IL-8 reveal a single class of IL-8 binding sites in HEL cells with a K_d of 7.4 nM and a receptor density of 12 818 binding sites/cell. In competition studies unlabeled IL-8, MGSA, MCP-1, and RANTES inhibit the binding of ^{125}I-IL-8 to HEL cells. This binding site was later identified as DARC[13,14].

Erythrocytes

^{125}I-IL-8 binds to chemokine receptors (CK) on erythrocyte membranes and to dodecyl β-maltoside solubilized CK receptors at a single class of high affinity binding sites in both cases with K_d values of 9.5 nM and 15.4 nM, respectively. This binding can be displaced by both α- and β-chemokines. This receptor is also known as DARC on erythrocytes[15].

Rabbit IL-8 receptors

^{125}I-IL-8 binds to rabbit IL-8 receptor 5B1a with a novel affinity binding profile of IL-8 >> NAP-2 > MGSA. The corresponding apparent K_i values for IL-8, NAP-2, and MGSA are 4, 120, and 320 nM, respectively.

Mouse IL-8 receptor homologue

IL-8 binds to the mouse homolog of the human IL-8 receptor with nanomolar affinity.

N51 receptor transfectants

The competition studies of ^{125}I-IL-8 with unlabeled IL-8 defines a single high-affinity site but competition with N51/KC defines the presence of both a high- and a low-affinity class of receptors. Hybrids of N51/IL-8 in which the N51/KC sequence between cysteines 2 and 3 (or first disulfide bond) is replaced by the corresponding sequence in IL-8 show IL-8-like properties. The IL-8 C-terminus deletion mutants are biologically inactive, but the hybrid molecules N51/IL-8 and IL-8/N51, in which the C-termini are exchanged, have biological activities similar to the wild-type molecules.

COS-7 cells transfected with IL-8 receptors

^{125}I-IL-8 binds to both type 1 and type 2 IL-8 receptors expressed in COS-7 cells. Binding to type 1 receptors is exclusive but binding to type 2 receptors is shared with MGSA and NAP-2.

F3R transfectants

IL-8 binds to COS-7 cells transfected with a cDNA clone, F3R, isolated from a neutrophil cDNA library with an affinity of 1.2 nM.

ECRF3 transfectants

IL-8 binds to viral chemokine receptor encoded by ECRF3 of *Herpesvirus saimiri* with a ligand specificity identical to that of IL-8RB.

Receptor cross-linking studies

Two proteins of approximately 70 kDa and 44 kDa (p70 and p44) are specifically cross-linked with labeled IL-8. Unlabeled IL-8 fully inhibits this cross-linking and the binding of labeled NAP-2 to the high-affinity sites on neutrophils or neutrophil membranes. Treatment of membranes with digitonin results in the preferential solubilization of a single receptor species, corresponding to p44, that binds IL-8 with high affinity ($K_d = 0.4$ nM). Exposure of neutrophil membranes to 100 μM GTPγS leads to a 75-fold increase of the K_d in approximately 60% of the receptors[16].

IL-8, but not NAP-2, can be cross-linked to dimers when bound to IL-8 receptor B, suggesting that IL-8 binds as a dimer and oligomer to IL-8 receptor.

Mutants of IL-8

Scanning mutagenesis of IL-8 identified the requirement of a cluster of residues at the N-terminus including an ELR motif (amino acids 4, 5 and 6) for receptor binding as well as functional activity of IL-8[17].

IL-8(4–72), a truncated form of IL-8 (IL-8(6–72)), and IL-8 AAR(7–72) with N-terminal Ala^4-Ala^5 instead of Glu^4-Leu^5, are potent antagonists of IL-8 binding, exocytosis (IC$_{50}$ 0.3 μM), as well as chemotaxis and the respiratory burst.

Mutation of IL-8 at Leu^{25} to Tyr introduces a novel monocyte chemoattractant activity into IL-8 and the mutant can displace ^{125}I-MIP-1α binding to CC-CKR-1 receptor with an affinity only 12-fold less than MIP-1α.

Replacement of rabbit IL-8 at His^{13} and Thr^{15} with Tyr^{13} and Lys^{15} of the human molecule converts the low-affinity binding of the rabbit IL-8 to the high-affinity binding of human IL-8. As a corollary, replacement of the Tyr^{13} and Lys^{15} of the human IL-8 with His^{13} and Thr^{15} of the rabbit IL-8 reduces the binding activity of this mutated human IL-8 by 200-fold.

Cross-desensitization

IL-8 pretreatment of neutrophils causes homologous desensitization (i.e., neutrophils are desensitized for subsequent stimulation with IL-8) as well as heterologous desensitization for subsequent stimulation with NAP-2 and MGSA. By contrast, following primary stimulation with NAP-2 or MGSA/GRO, responses to IL-8 are only moderately attenuated, supporting the existence of IL-8 receptors which bind NAP-2 or MGSA/GRO with low affinity[18].

IL-8 pretreament of neutrophils desensitizes their subsequent stimulation by ENA-78, but pretreatment with ENA-78 does not cause desensitization of IL-8 stimulation.

Stimulation of neutrophils with IL-8 resulted in desensitization toward a subsequent challenge with NAP-2 or MGSA/GRO as shown by the rise in cytosolic free calcium. By contrast, following primary stimulation with NAP-2 or MGSA/GRO, responses to IL-8 were only moderately attenuated.

Gene structure

Human

Canine

Gene location

The human IL-8 gene maps to 4q12–q21, which is the same chromosomal location as that of three other members of the PF4 gene superfamily, PF4, IP-10, and MGSA.

Protein structure

IL-8 dimer is a globular structure made up of six-stranded antiparallel β-sheet, three strands contributed by each subunit, on top of which lie two antiparallel helices separated by approximately 14 Å, and the symmetry axis is located between residues 26 and 26' [19].

Although IL-8 forms dimers in solution, monomeric IL-8 is also functionally active in the induction of chemotaxis in neutrophils[20]. Monomer–dimer equilibrium studies of IL-8 using analytical ultracentrifugation and titration microcalorimetry indicate that it dissociates readily to monomers with an equilibrium dissociation constant of 18 μM at 37°C.

Studies with hybrid proteins between functional IL-8 and non-functional IP-10 indicate that Gly^{31} and Pro^{32}, as well as the N-terminal region from IL-8 are required to convert IP-10 into a fully functional protein, suggesting that these elements are critical for IL-8 activity. Both disulfide bridges, linking residue 7 to 34 and residue 9 to 50, are critical for function. The presence of residues within the 10–15 region and the 17–22 region is critical for functionality of IL-8. Tyr^{13}, Phe^{17}, and Phe^{21} are involved in aromatic interactions in the IL-8 structure. Except for Cys^{50} no evidence is known for the role for the 36–72 region, including the C-terminal α-helix, in receptor binding.

Presence or absence of the ELR motif determines the angiogenic or angiostatic effect of IL-8. Mutant IL-8 proteins lacking the ELR motif demonstrate potent angiostatic effects in the presence of either ELR-containing α-chemokines or basic FGF[21].

Chemically synthesized analogs of IL-8 corresponding to the less abundant natural forms 3–72 and 4–72 have 2–5-fold higher potencies than IL-8, whereas 77-residue IL-8 is 2-fold less potent. N-terminal residues 4, 5, and 6 (E, L, and R) are absolutely essential for IL-8 activity and receptor binding. Analog 5–72 is 80-fold less potent than IL-8 in an elastase assay, but is only slightly less potent in stimulating chemotaxis. Analog 6–72 lacks all biological activities but has detectable receptor-binding activity. Comparison of analogs shortened at the C-terminus shows that potency is progressively reduced as the C-terminal residues are excluded. However, activity is retained in an analog (1–51) with the entire C-terminal α-helix and β-turn missing. A peptide corresponding to the C-terminal 22 residues, although inactive alone, synergizes with the 1–51 analog in stimulating elastase release[22].

Amino acid sequence

Human

```
1   MTSKLAVALL AAFLISAALC EGAVLPRSAK ELRCQCIKTY SKPFHPKFIK
51  ELRVIESGPH CANTEIIVKL SDGRELCLDP KENWVQRVVE KFLKRAENS
```

Dog

```
1    MTSKLAVALL AAFVLSAALC EAAVLSRVSS ELRCQCIKTH STPFHPKYIK
51   ELRVIDSGPH CENSEIIVKL FNGNEVCLDP KEKWVQKVVQ IFLKKAEKQD
101  P
```

Red-crowned mangabey

```
1   MTSKLAVALL AAFLLSAALC EGAVLPRSAK ELRCLCIKTY SKPFHPKFIK
51  ELRVIESGPH CVNTEIIVKL SDGRELCLDP KEPWVQRVVE KFLKRAESQN S
```

Pig

```
1    MTSKLAVAFL AVFLLSAALC EAAVLARVSA ELRCQCINTH STPFHPKFIK
51   ELRVIESGPH CENSEIIVKL VNGKEVCLDP KEKWVQKVVQ IFLKRTEKQQ
101  QQQ
```

Rabbit

```
1    MNSKLAVALL ATFLLSLTLC EAAVLTRIGT ELRCQCIKTH STPFHPKFIK
51   ELRVIESGPH CANSEIIVKL VDGRELCLDP KEKWVQKVVQ IFLKRAEQQE
101  S
```

Sheep

```
1    MTSKLAVALL AAFLLSAALC EAAVLSRMST ELRCQCIKTH STPFHPKFIK
51   ELRVIESGPH CENSEIIVKL TNGKEVCLDP KEKWVQKVVQ AFLKRAEKQD
101  P
```

Database accession numbers

	GenBank	SwissProt	PIR	MIM	Ref
Human	M17017	P10145	A37037	146930	23
	M26383		S03975		
	M28130		S04216		
	M14283				
Dog	D28772	P41324			24
Mangabey	G644796	P26894	A26774		
Pig	M86923				25
Rabbit	M57439	P19874	S13052		26
Sheep	X78306	P36925	S42496		27

References

[1] Matsushima, K. and Oppenheim, J.J. (1989) *Cytokine* 1, 2–13.

[2] Zhang, Y. et al. (1995) *J. Clin. Invest.* 95, 586–592.

[3] Xu, L. et al. (1995) *J. Leukoc. Biol.* 57, 335–342.

[4] Stein, B. and Baldwin, A.S. (1993) *Mol. Cell. Biol.* 13, 7191–7198.

[5] Parry, G.C. and Mackman, N. (1994) *J. Biol. Chem.* 269, 20823–20825.

[6] Oliveira, I.C. et al. (1994) *Mol. Cell. Biol.* 14, 5300–5308.

[7] Simonet, W.S. et al. (1994) *J. Clin. Invest.* 94, 1310–1319.

[8] Laterveer, L. et al. (1995) *Blood* 85, 2269–2275.

[9] Barnett, M.L. et al. (1993) *Biochim. Biophys. Acta* 1177, 275–282.

[10] Richardson, R.M. et al. (1995) *Biochemistry* 34, 14193–14201.

[11] Moser, B.C. et al. (1991) *J. Biol. Chem.* 266, 10666–10671.

[12] Krieger, M. et al. (1992) *J. Immunol.* 149, 2662–2667.

[13] Horuk, R. et al. (1993) *Biochemistry* 32, 5733–5738.

[14] Horuk, R. et al. (1994) *J. Biol. Chem.* 269, 17730–17733.

[15] Horuk, R. et al. (1993) *Science* 261, 1182–1184.

[16] Schumacher, C. et al. (1992) *Proc. Natl. Acad. Sci. USA* 89, 10542–10546.

[17] Hebert, C.A. et al. (1991) *J. Biol. Chem.* 266, 18989–94.

[18] Moser, B. et al. (1991) *J. Biol. Chem.* 266, 10666–10671.

[19] Clore, G.M. and Gronenborn, A.M. (1995) *FASEB J.* 9, 57–62.

[20] Rajarathnam, K. et al. (1994) *Science* 264, 90–92.

[21] Strieter, R.M. et al. (1995) *J. Biol. Chem.* 270, 27348–27357.

[22] Clark-Lewis, I. et al. (1991) *J. Biol. Chem.* 266, 23128–23134.

[23] Matsushima, K. et al. (1988) *J. Exp. Med.* 167, 1883–1893.

[24] Ishikawa, J. et al. (1993) *Gene* 131, 305–306.

[25] Lin, G. et al. (1994) *J. Biol. Chem.* 269, 77–85.

[26] Yoshimura, T. and Yukhi, N. (1991) *J. Immunol.* 146, 3483–3488.

[27] Legastelois, I. et al. (1994) *Gene* 150, 367–369.

Alternative names

GRO-α, neutrophil activating protein 3 (NAP-3), KC (murine), N51 (murine), cytokine-induced neutrophil chemoattractant (CINC) (rat), rabbit permeability factor 2 (RPF-2)

Family

α family (CXC family)

Molecule

MGSA was originally found as a mitogenic polypeptide secreted by human melanoma cells. The structural similarity of MGSA to other members of the chemokine family resulted in the investigation of its role in the inflammatory process. The MGSA/GRO-α gene product has potent chemotactic, growth regulatory, and transformative functions[1].

Tissue sources

Monocytes, human nasal and bronchial epithelium, bronchoalveolar macrophages, rat anterior pituitary cells, neutrophils, endothelial cells, epidermal keratinocytes, and different tumor cells such as melanoma, benign intradermal, and dysplastic nevi, MeWo melanoma cells, human mast cell line HMC-1.

Target cells

Neutrophils, lymphocytes, monocytes, and epidermal melanocytes.

Physicochemical properties

Property	Human	Mouse	Hamster	Rat	Rabbit
pI	~9	~9	~9	~9	?
Signal	1–34	1–24	1–28	1–24	?
Amino acids					
Precursor	107	96	101	96	71
Mature	72	72	72	71	71
Disulfide bonds	a.a. 43–69	33–59	37–63	33–59	7–33
	a.a. 45–85	35–75	39–79	35–75	9–49
Glycosylation sites	0	0	0	0	0
Molecular weight					
Predicted	11301	10254	10893	10249	7713

Transcription factors

NF-κB p50/p65 heterodimer in normal retinal pigment epithelial (RPE) cells but not melanoma cells plays an important role in IL-1- and TNF-α-enhanced gene transcription of MGSA/GRO[2].

Basal and cytokine-induced expression of MGSA/GRO-α requires at least three transcription factors, Sp1, NF-κB, and HMGI(Y) in the melanoma cell line Hs294T.

The *CINC* gene contains a binding site in the promoter region for NF-κB.

The LPS-responsive region in the KC gene promoter is located between residues −104 and +30. This region contains two κB sequence motifs. The first motif (position −70 to −59, κB1) is highly conserved in all three human *GRO* genes and in the mouse *Mip*-2 gene. The second κB motif (position −89 to −78, κB2) is conserved between the mouse and the rat *KC* genes. Both κB sites are essential for inducibility by LPS in RAW264.7 cells and by TNF-α in NIH 3T3 fibroblasts. Although both κB1 and κB2 sequences are able to bind members of the Rel homology family, including NF-κB1 (p50), RelA (p65), and c-Rel, the κB1 site binds these factors with higher affinity and functions more effectively than the κB2 site in a heterologous promoter[3].

Regulation of expression

Stimulus	Cell type	Response
Stress		
Ultraviolet B radiation	Keratinocyte cell line, A431	Up-regulation of MGSA expression, inhibitable by IL-1 receptor antagonist
Bacteria		
LPS	Endothelial cells	Induction of MGSA gene expression
Bacterial endotoxin	Rat alveolar macrophages	Induction of CINC mRNA expression
Cytokines		
IL-1	RPE cells	10–20-fold increase in MGSA gene transcription
	Human melanoma cell line, Hs294T cells	2-fold increase in MGSA gene transcription
	Synovial cells	Superinduction of MGSA gene after inhibition of protein synthesis
IL-1-β	Human gingival fibroblasts	Induction of MGSA gene expression
TNF-α	Bronchial epithelial cells	100-fold up-regulation of GRO-α
IL-1 and TNF-α	RPE and Hs294 cells	Post-transcriptional regulation of MGSA
	Endothelial cells	Induction of MGSA gene expression
	Human synovial cells and fibroblasts	Induction of MGSA expression
	Normal rat kidney fibroblasts, NRK-49F cells	Induction of CINC expression
PDGF	Human melanoma cell line, Hs294T	Induction of MGSA mRNA expression
IFN-γ	Mouse peritoneal macrophages	Inhibition of LPS-inducible KC gene expression
EGF	Growth-arrested mammary epithelial cells	High levels of synthesis of MGSA within 1 h

Stimulus	Cell type	Response
Others		
Thrombin	Human endothelial cells	MGSA mRNA expression
MM-LDL	Human and rabbit aortic endothelial cells	Induction of MGSA mRNA expression
	Human endothelial cells	Increase in a surface-associated protein that binds to antibody against GRO despite low levels of GRO in the medium
Glucocorticoids	Macrophages	Inhibition of serum-induced expression of KC
		Inhibition of CINC production induced by TNF or IL-1 with potencies dexamethasone > prednisolone > hydrocortisone
Serum	RPE cells	Induction of expression of GRO-α but not GRO-β
PMA	Monocytes	Induction of MGSA/GRO gene
Adrenomedullin	Rat alveolar macrophages	Inhibition of CINC secretion

[a] Higher constitutive expression is seen in melanoma cells than other cell types such as normal retinal pigment epithelial cells.

Expression in disease

Human studies

Freshly isolated human synovial fibroblasts from patients with rheumatoid arthritis (RA) contain elevated levels of GRO-α[4]. Synovial fluid (SF) and plasma of patients with RA but not with osteoarthritis express high levels of MGSA/GRO-α protein and account for about 30% of neutrophil chemotactic activity of RA SF. In contrast, the majority of synovial fibroblast cell lines derived from osteoarthritic or noninflammatory synovia show a relative increase in the constitutive expression of GRO-α over synovial fibroblasts obtained from rheumatoid synovia.

Elevated expression of GRO-α is found in human colonic tumors.

MGSA is found in the cytoplasmic granules in the melanosomes in metastatic melanoma patients[5].

More than 150-fold elevation of MGSA is found in lesional psoriasis tissues when compared to normal heel callus. MGSA is the second major chemoattractant for neutrophils purified from psoriatic scales by multistep HPLC.

Animal studies

The mRNA encoding KC is found in spinal cord 2 days before clinical signs are apparent in murine EAE.

Spontaneous release of KC/GRO protein from the hepatocytes of chronically ethanol-fed rats is markedly enhanced.

Rat CINC is one of the two chemotactic factors isolated from rat inflammatory exudate induced by a subcutaneous injection of LPS in a carboxymethyl-cellulose suspension.

The mRNA for the chemokine CINC is induced in the kidney, and the corresponding protein is elaborated by isolated inflamed glomeruli in the rat model of antiglomerular basement membrane (GBM) nephritis.

Elevated expression of CINC mRNA is found in the ipsilateral cerebral cortex of rats subjected to focal cerebral ischemia induced by middle cerebral artery occlusion.

In vivo studies

Angiogenic effect is shown by corneal neovascularization in vivo[6].

Intradermal injection of human MGSA into rats leads to massive infiltration of neutrophils.

Intratracheal instillation of KC induces dose-dependent PMN influx into the airspaces. A neutralizing anti-KC antibody reduces the chemotactic activity of rat bronchoalveolar lavage fluid collected after the intratracheal administration of LPS and inhibits PMN accumulation within the lungs in response to an intratracheal challenge of LPS.

In vitro biological effects

Erythrocytes

Inhibits erythrocyte invasion by malarial parasite Plasmodium.

Fibroblasts

MGSA induces a decrease in the expression of interstitial collagens in rheumatoid synovial fibroblasts at 0.6–6 nM range.

Neutrophils[7,8]

Calcium mobilization
Chemotaxis (2 ng/ml)
Exocytosis of elastase
Respiratory burst
MGSA primes human neutrophils for fMLP induced superoxide production and up-regulation of fMLP receptors
KC induces chemotaxis, respiratory burst, and increased expression of CD11b/CD18.

Basophils

Chemotaxis
Intracellular calcium changes

Pituitary cells

CINC/GRO stimulates IL-6 secretion by rat posterior pituitary cells.

CINC/GRO increases the secretion of PRL (10–100 ng/ml) and suppresses basal LH and FSH secretions.

Cell lines

MGSA induces cellular proliferation in human melanoma cell line Hs294T[9].

MGSA produces rapid Ca^{2+} flux in U937 cells.

Transfection of human MGSA into mouse melanoma cells leads to increased ability to form large colonies in soft agar and increased ability to form tumors when injected into nude mice.

Intracellular signaling

MGSA binding to purified human neutrophil plasma membranes leads to stimulation of high affinity GTPase and GTP hydrolysis.

MGSA binding to the placental cell line $3A_E$ P-3 transfected with IL-8RB causes tyrosine phosphorylation of p130/cas antigen (crk-associated protein) associated with the membrane and p70 associated with cytosol[10].

MGSA binding to IL-8RB transfected into nonhematopoietic cells results in phosphorylation of the receptor on serine residues and subsequent degradation upon prolonged exposure.

Receptor binding characteristics

Cell/receptor	Radiolabeled ligand	Competing ligand	Affinity/receptor[a]
Neutrophils	IL-8	MGSA	70% HA (K_d: 0.14 nM) 30% LA (K_d: 130 nM)
IL-8RA	MGSA	MGSA	400 nM
IL-8RB	MGSA	MGSA	HA
DARC	MGSA	MGSA	HA
	IL-8	MGSA	HA
Duffy antigen on erythrocytes	MGSA	MGSA	HA
ECRF3	MGSA	MGSA	HA
Hs294T cells	MGSA	MGSA	3.9- 4.25 nM 52 960–67 758
U937 cells	MGSA	MGSA	3.2 nM
	MGSA	IL-8	~5 nM
Chimeric DARCe1/IL-8RB	MGSA	MGSA	LA
	IL-8	MGSA	LA
	MCP-1	MGSA	LA
	RANTES	MGSA	LA

Cell/receptor	Radiolabeled ligand	Competing ligand	Affinity/receptor
Rabbit homolog of IL-8RB (5B1a)	MGSA	MGSA	300 nM
Neutrophils	KC	KC	HA
	IL-8	KC	HA
	MGSA	KC	HA
IL-8RB homolog	KC	KC	3 nM

[a] HA, high affinity; LA, low affinity.

Studies with mutants of MGSA or KC

MGSA binding to IL-8RB is dependent on the presence of the intact N-terminus. It is completely prevented by replacement of the ELR sequence with alanines.

Mutant MGSA (glu(6) → Ala) binds DARC but not IL-8RB.

C-terminal α-helix of MGSA (47–71) binds human melanoma cell line Hs294t in a non-ELR-dependent manner.

The competition studies of ^{125}I-IL-8 with N51/KC define the presence of both a high- and a low-affinity class of receptors. Hybrids of N51/IL-8 in which the N51/KC sequence between cysteines 2 and 3 (or first disulfide bond) is replaced by the corresponding sequence in IL-8 show IL-8-like properties, indicating that this region is important for specific receptor recognition. The N51 δIII C-terminus deletion mutant is biologically inactive, but the hybrid molecules N51/IL-8 III and IL-8/N51 III, in which the C-termini are exchanged have biological activities similar to that of the wild-type molecules, demonstrating that the presence of the C-terminus is essential.

Cross-desensitization

Stimulation of human neutrophils with IL-8 completely desensitizes them toward a subsequent challenge with MGSA. By contrast, following primary stimulation with MGSA, responses to IL-8 are only moderately attenuated[11].

Pretreatment of human neutrophils with IL-8 results in the subsequent desensitization to murine N51/KC. In contrast, pretreatment with N51/KC does not result in a loss of response to a subsequent treatment with IL-8.

Pretreatment of neutrophils with GRO-α desensitizes subsequent challenge with GRO-β and GRO-α.

Pretreatment of neutrophils with MGSA or ENA-78 cross-desensitizes their subsequent stimulation by either MGSA or ENA-78.

Gene structure

Gene location

The human *MGSA* gene is located on 4q21.

The mouse *Mgsa* gene is polymorphic and syntenic with the W, patch, rumpwhite, and recessive spotting loci on chromosome 5. The mouse gene is localized between *Afp* and *Gus* on chromosome 5[12].

Protein structure

The MGSA monomer contains an N-terminal loop, a three-stranded antiparallel β-sheet arranged in a 'Greek key' conformation, and a C-terminal α-helix (residues 58–69). Dimerization, which is apparent under the experimental conditions used (2 mM, pH 5.10, 30°C), results in a six-stranded antiparallel β-sheet and a pair of helices with 2-fold symmetry. While the basic fold is similar to that seen for IL-8 there are differences in the ELR motif (residues 6–8), and the turn involving residues 31–36, which is linked to the N-terminal region through the 9–35 disulfide bond. The most significant differences are in the N-terminal loop (residues 12–23). At the quaternary (dimer) level the difference between IL-8 and MGSA results from differing angles between the β-strands which form the dimer interface, and is manifested as a different interhelical separation (distance of closest approach between the two helices is 15.3 Å in the IL-8 NMR structure and 11.7 (±0.4) Å in the MGSA structure). In IL-8, all the corresponding regions have been shown to be required for receptor binding[13].

Amino acid sequence

MGSA Human

```
1    MARAALSAAP SNPRLLRVAL LLLLLVAAGR RAAGASVATE LRCQCLQTLQ
51   GIHPKNIQSV NVKSPGPHCA QTEVIATLKN GRKACLNPAS PIVKKIIEKM
101  LNSDKSN
```

GRO-α Rat

```
1    MVSATRSLLC AALPVLATSR QATGAPVANE LRCQCLQTVA GIHFKNIQSL
51   KVMPPGPHCT QTEVIATLKN GREACLDPEA PMVQKIVQKM LKGVPK
```

GRO-α Hamster

```
1    MAPATRSLLR APLLLLLLLL ATSRLATGAP VANELRCQCL QTMTGVHLKN
51   IQSLKVTPPG PHCTQTEVIA TLKNGQEACL NPEAPMVQKI VQKMLKSGIR
101  K
```

GRO-α Mouse

```
1    MIPATRSLLC AALLLLATSR LATGAPIANE LRCQCLQTMA GIHLKNIQSL
51   KVLPSGPHCT TEVIATLKN GREACLDPEA PLVQKIVQKM LKGVPK
```

GRO-α Rabbit

```
 1   ALTELRCQCL QTVQGIHLKN IQNLKVLSPG PHCAQTEVIA TLKSGQEACR
51   NPAAPMVKKF LQKRLSNGNS S
```

Database accession numbers

	GenBank	SwissProt	PIR	MIM	Ref
Human	G34519	P09341	A28414	155730	[14]
	J03561		S03976		
	G306806		S13669		
	X12510				
	G34622				
	X54489				
Mouse	J04596	P12850	A32954		[15]
(KC)	G201043		JH0081		
Hamster	J03560	P09340	B28414		[16]
Rabbit	G309960	P30782	S17507		[17]
	L19157				
Rat	D11445	P14095	A34481		

References

[1] Richmond, A. (1991) *Semin. Dermatol.* 10, 246–255.

[2] Shattuck, R., Wood, L.D. et al. (1994) *Mol. Cell. Biol.* 14, 791–802.

[3] Ohmori, Y. et al. (1995) *J. Immunol.* 155, 3593–3600.

[4] Hosaka, S. et al. (1994) *Clin. Exp. Immunol.* 97, 451–457.

[5] Priest, J.H. et al. (1988) *Cancer Genet. Cytogenet.* 35, 253–262.

[6] Strieter, R.M. et al. (1995) *J. Biol. Chem.* 270, 27348–27357.

[7] Moser, B. et al. (1990) *J. Exp. Med.* 171, 1797–1802.

[8] Geiser, T. et al. (1993) *J. Biol. Chem.* 268, 15419–15424.

[9] Horuk, R. et al. (1993) *J. Biol. Chem.* 268, 541–546.

[10] Schraw, W. and Richmond, A. (1995) *Biochemistry* 34, 13760–13767.

[11] Moser, B. et al. (1991) *J. Biol. Chem.* 266, 10666–10671.

[12] Balentien, E. et al. (1990) *Biochemistry* 29, 10225–10233.

[13] Kim, K.S. et al. (1994) *J. Biol. Chem.* 269, 32909–32915.

[14] Baker, N.E. et al. (1990) *Nucleic Acids Res.* 18, 6543–6456.

[15] Oquendo, P. et al. (1989) *J. Biol. Chem.* 264, 4133–4137.

[16] Anisowicz, A. et al. (1987) *Proc. Natl. Acad. Sci. USA* 84, 7188–7192.

[17] Johnson, M.C. et al. (1994) *Gene* 151, 337–338.

[18] Konishi, K. et al. (1993) *Gene* 126, 285–286.

Family

α family (CXC family)

Molecule

PF4 is a heat stable heparin binding protein, stored in the α-granules of platelets along with fibronectin, fibrinogen, thrombospondin, von Willebrand factor, and β-TG[1] and secreted from platelets following stimulation and aggregation. PF4 is also described as a marker for megakaryocyte differentiation and as a negative autocrine regulator of human megakaryocytopoiesis. Another related protein, low-affinity platelet factor 4 (LA-PF4), bearing 50% homology to PF4 is also described. LA-PF4, unlike PF4, is an active mitogenic and chemotactic agent and does not bind to heparin[2].

PF4 is associated with a carrier molecule, proteoglycan, *in vivo*. This proteoglycan is 53 kD and consists of 32% uronic acid, 31% galactosamine, 6.1% sulfate, and 9.9% protein. The interaction between this proteoglycan carrier and PF4 is strongly dependent on ionic strength; 0.3 M NaCl is required to dissociate the proteoglycan PF4 complex[3].

Tissue sources

Platelets, megakaryocytes

Target cells

Fibroblasts, platelets, adrenal microvascular pericytes, mast cells, basophils, megakaryocytes

Physicochemical properties

Property	Human	Rat	Ovine	Porcine
pI	8.7	9.9	8.3	9.6
Signal	1–31	1–29	?	?
Amino acids				
Precursor	101	105	85	90
Mature	70	105	85	90
Disulfide bonds	a.a. 41–67	44–71	25–51	25–51
	a.a. 43–83	46–87	27–67	27–67
N-Glycosylation sites	0	1 (a.a. 31)	0	1 (a.a. 8)
Molecular weight	10 845	11 286	9129	9644

Transcription factors

Transient expression experiments with rat bone marrow cells and other cell lines reveal a complex interplay between a core promoter domain from −97 to the transcriptional start site and an enhancer/silencer domain from −448 to −112. The core promoter contains a GATA site at −31 to −28 whose mutation to TATA or AATA decreases tissue specificity and moderately affects expression in megakaryocytes as well as a positively acting subdomain from

−97 to −83 whose removal decreases overall transcription without affecting tissue specificity. The enhancer/silencer domain possesses three positively acting subdomains from −380 to −362, −270 to −257, and −137 to −120 as well as a negatively acting subdomain at −184 to −151 which is able to reduce overall transcription but has no effect on tissue specificity. The subdomain from −380 to −362 is most critical in restricting gene expression driven either by the PF4 promoter or by a heterologous promoter in the megakaryocytic lineage. The subdomains from −270 to −257 and −137 to −120 function together with the subdomain from −380 to −362 to somewhat increase tissue specificity. Simultaneous mutation of the GATA site and deletion of either the whole enhancer/silencer domain or the subdomain from −380 to −362 or −137 to −120 reduce transcription in megakaryocytes by 10- to 30-fold. It is proposed that megakaryocyte-specific enhancer/silencer domain and the GATA site are responsible for high-level expression of the PF4 gene in a lineage-specific manner[4].

Regulation of expression

A clone of the human cell line CMK that spontaneously expresses megakaryocytic characteristics, named CMK 115, expresses PF4 mRNA which can be increased by the addition of TPA.

Regulation of release

Platelets release PF4 following stimulation with ADP and collagen along with serotonin

Thrombin causes release of PF4, but at concentrations of thrombin about 100 times more than that which is required for fibrinogen cleavage

Platelet activating factor causes the release of PF4 from platelets

In vivo studies

Human studies

Elevated PF4 levels are seen during cardiopulmonary bypass surgery, in arterial thrombosis, following surgery, in acute myocardial infarction, during acute infections, and in inflammatory states.

PF4 levels are normal in immune thrombocytopenia, disseminated malignancy, severe hepatic and renal disease, and chronic arterial disease.

Patients with short platelet survival time have a high plasma concentration of PF4.

Elevated urinary excretion and lower circulating levels of PF4 are seen in a majority of patients with myeloproliferative disorders such as myelofibrosis.

Intravenous injection of heparin (100 U/kg) into normal human volunteers results in an increase of PF4 levels in platelet-poor plasma from a mean value of 18.1 ng/ml before injection to 257.9 ng/ml at 5 min after injection. This increase is due to release of PF4 bound to endothelial cells.

Intravenous injection of human or mouse serum or platelet material secreted from appropriately stimulated platelets containing PF4 together with antigen alleviates the immunosuppression and restores plaque-forming cells to unsuppressed numbers in SJL/J mice induced by injection of irradiated lymphoma cells or in (CB6) F1 mice induced by injection of Con A. However, this effect requires coversion of PF4 to an active form by a platelet protease (PF-101).

Animal studies

The most signficant effect of PF4 *in vivo* is its anti-angiogenic effect. The angiostatic effect is evidenced by inhibition of angiogenesis induced by ELR containing chemokines such as IL-8 and MGSA/GRO-α[5]. PF4 inhibits angiogenesis in the chicken chorioallantoic membrane.

A truncation mutant of PF4, termed PF4-241 (rPF4-241), does not bind heparin but retains the ability to suppress the growth of tumors in mice. Daily intralesional injections of rPF4-241 significantly inhibit the growth of the B-16 melanoma in syngeneic mice without direct inhibitory effects on B-16 cell growth *in vitro*. Similar antitumor effects were observed with the human colon carcinoma, HCT-116, grown in nude mice, indicating that the inhibitory activity was neither tumor-type specific nor T-cell dependent[6].

PF4 alleviates Con A-induced immunosuppression in mice. PF4 prevents the induction of Con A-induced suppressor cells *in vitro* but not the function of established Con A suppressor cells. In DTH reaction to sheep red blood cells (SRBCs) PF4 enhances the magnitude of the swelling following SRBC challenge.

Expression of mRNA is found in the glomeruli in a murine model of anti-glomerular basement membrane antibody induced glomerulonephritis.

Infusion of prostacyclin (PGI$_2$) (100 or 300 ng/kg/min) into monkeys for 15 min results in a decrease of plasma levels of PF4 by 40%–60%.

In vitro biological effects

General effects

Neutralization of the anti-coagulant and anti-factor Xa activities of heparin
Inhibition of collagenase
Stimulation of human leukocyte elastase activity

Progenitor cells

Suppression of colony formation of granulocyte-macrophage, erythroid, and multipotential progenitor cells.

Vascular cells

Stimulation of migration of bovine adrenal microvascular pericytes (1 µg/ml)
Inhibition of endothelial cell proliferation

Mast cells

C-terminal dodecapeptide of PF4(59–70) causes the release of histamine and concurrently generates leukotrienes B4, C4, and D4 in isolated dog mastocytoma cells.

Basophils

Stimulation by PF4 and the synthetic substituent peptide PF4(59–70) causes the release of histamine noncytotoxically from human basophils (10^{-7} M PF4 and 10^{-5} M PF4(59–70)).

Megakaryocytes

Inhibition of megakaryocyte colony formation with no effect on either myeloid or erythroid colony formation (25 µg/ml). Specific effects on mixed colony forming unit-megakaryocytes (mCFU-MK), the burst forming unit-megakaryocytes (BFU-MK), and the colony forming unit-megakaryocytes (CFU-MK), are seen at a PF4 concentration of 2.5 µg/ml or greater[7].

Fibroblasts

Chemotaxis (200 ng/ml)

Cell lines

3T3 cell lines adhere to PF4 adsorbed on to hyaluronate-binding substrata. PF4 stimulates formation of broad convex lamellae but not tapered cell processes fibers during the spreading response. PF4-mediated responses are blocked by treating the PF4-adsorbed substratum with heparin (but not chondroitin sulfate), or with *Flavobacter heparinum* heparinase (but not chondroitinase ABC). Microtubular networks reorganize in cells on PF4 but stress fibers are not seen in these cells[3].

A significant inhibition of HEL colony growth and thymidine incorporation is seen at PF4 doses of 1 µg/ml and 0.5 µg/ml, respectively. The inhibitory effect of PF4 is abrogated by the addition of heparin (5–10 µg/ml) (PF-48). In HEL cells, PF4 inhibits colony formation by impeding cell maturation through inducing expression of c-*myc* and c-*myb*. PF4 also inhibits the expression of factor V mRNA (100 ng/ml).

Receptor binding characteristics

Binding of ^{125}I-PF4 to HEL cells reaches equilibrium within 20–30 min with a dissociation constant of 1.3×10^{-10} M and B_{max} of 6.3 pmol/10^5 cells and is inhibited by an excess of unlabeled PF4, β-TG, and heparin[8].

Gene structure

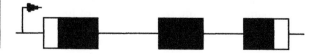

The transcriptional start site is located 73 bp upstream of the translational start codon. The 5′ noncoding region of the gene also exhibits a sequence homologous to the TATA box at −31, as well as a series of direct and inverted repeat sequences and a cluster of 26 T residues at −155 to −218. This latter domain is involved in regulating PF4 gene expression during megakaryocyto-poiesis[16].

Another highly homologous gene to that of PF4, PF4alt, is also reported. PF4 and PF4alt are non-allelic genes: the human *PF4* gene is encoded on a 10 kb EcoRI fragment, while *PF4alt* is encoded in a polymorphic 3 or 5 kb EcoRI fragment. Compared with *PF4*, this gene has 14% DNA and 38% amino acid divergence in the signal peptide region, and 2.6% DNA and 4.3% amino acid divergence in the coding region of the mature protein. *PF4alt* contains three amino acid substitutions (P58−L, K66−E, and L67−H) near the C-terminus, in a region known to be critical for PF4 function. The pyrimidine-rich region present in the rodent *PF4* gene is absent from the *PF4alt* gene.

Gene location

The *PF4* gene, like the *β-TG* gene, is located on chromosome 4. It is, there-fore, possible that the *β-TG* and *PF4* genes form a single genomic locus on chromosome 4 whose members become coordinately activated during megakaryocyte differentiation. The *PF4* gene, like most of the CXC α-chemo-kine genes, is located on a single 700 kb SfiI fragment localized to chromo-some bands 4q12–q13. Analysis of λ genomic clones demonstrates that the *PF4* and *β-TG* genes are separated by less than 7 kb, and the *β-TG2* and *PF4alt* genes by approximately 5 kb. Within each *β-TG/PF4* duplication, the *β-TG*-like gene is upstream of its linked *PF4*-like gene[9].

Protein structure

Human PF4 consists of 4 identical subunits containing 70 amino acids, each with a molecular weight of 7756. From consideration of the homology with β-TG, disulfide bonds between residues 10 and 36 and between residues 12 and 52 can be inferred. The secondary structure of the PF4 subunit, from N- to C-termini, consists of an extended loop, three strands of antiparallel β-sheet arranged in a Greek key, and one α-helix. The tetramer contains two extended, six-stranded β-sheets, each formed by two subunits, which are arranged back-to-back to form a 'β-bilayer' structure with two buried salt bridges sandwiched in the middle. The C-terminal α-helices, which contain lysine residues that are thought to be intimately involved in binding heparin, are arranged as antiparallel pairs on the surface of each extended β-sheet[10].

As a function of protein concentration, proton NMR spectra of human PF4 differ. Concentration-dependent NMR spectral changes are related to PF4 aggregation, with tetramers being the largest aggregates formed. There is a bimolecular mechanism of aggregation which proceeds from monomers to tetramers through a dimer intermediate. At pH 4, equilibrium constants for dimer association (K_D) and tetramer association (K_T) have the same value and resonances associated with all three aggregate states are observed. Lowering the pH shifts the equilibrium to the monomer state, while raising the pH shifts the equilibrium to dimer and tetramer states. Analysis of the pH dependence of K_D and K_T suggests that electrostatic interactions, probably arising from Glu/Asp and Lys/Arg side chains, play a role in the binding process. Increasing the solvent ionic strength stabilizes the tetramer state, especially at low pH, suggesting that intersubunit, repulsive electrostatic interactions probably between/among cationic side chains (Lys/Arg) attenuate the aggregation process. Information based primarily on histidine pK_a values and photo-CIDNP ^{1}H NMR data suggests that Tyr-60 and His-I, but not His-II, are significantly affected by the aggregation process. For dimer and tetramer equilibria at 30°C, unimolecular dissociation rate constants are $35 \pm 10\,\text{s}^{-1}$ for dimer dissociation and $6 \pm 2\,\text{s}^{-1}$ for tetramer dissociation[11].

Unlike PF4, LA-PF4 monomers are highly favored over dimers and tetramers. The biologically active state for PF4 is as a tetramer, while for LA-PF4 it is a monomer. Quaternary structure may, therefore, account for strong heparin binding in PF4, most likely by presenting a more favorable structural matrix for effective glycosaminoglycan interactions.

Amino acid sequence

Human

```
1    MSSAAGFCAS RPGLLFLGLL LLPLVVAFAS AEAEEDGDLQ CLCVKTTSQV
51   RPRHITSLEV IKAGPHCPTA QLIATLKNGR KICLDLQAPL YKKIIKKLLE S
```

Rat

```
1    MSAAAVFRGL RPSPELLLLG LLLLPAVVAV TRASPEESDG DLSCVCVKTS
51   SSRIHLKRIT SLEVIKAGPH CAVPQLIATL KNGSKICLDR QVPLYKKIIK
     KLLES
```

Sheep

```
1    XSSLPAASVS LPADSEGGEE EDLQCVCLKT TSGIHPRHIS SLEVIGAGLH
51   CPSPQLIATL KTGRKICLDQ QNPLYKKIIK RLLKN
```

Ovine PF4 shows 78% homology with bovine PF4, 76% with porcine PF4, 71% homology with human PF4, and 61% with rat PF4. The heparin binding site of ovine PF4 localized in the C-terminal region of the molecule is identified as LYKKIIKRLL.

Pig

```
1    QEWSLPGTRV PPPADPEGGD ANLRCVCVKT ISGVSPKHIS SLEVIGAGPH
51   CPSPQLIATL KKGHKICLDP QNLLYKKIIK KLLKSQLLTA
```

Database accession numbers

	GenBank	SwissProt	PIR	MIM	Ref
Human	M25897	P02776	A03241 A60161 S07159	173460	12
Rat	M15254	P06765	A26774		14
Sheep		P30035			15
Pig		P30034			13

References

1 Wencel-Drake, J.D. et al. (1986) *Blood* 68, 244–249.
2 Mayo, K.H. (1991) *Biochemistry* 30, 925–934.
3 Huang, J.S. et al. (1982) *J. Biol. Chem.* 257, 11546–11550.
4 Ravid, K. et al. (1991) *Mol. Cell. Biol.* 11, 6116–6127.
5 Strieter, R.M. et al. (1995) *J. Biol. Chem.* 270, 27348–27357
6 Maione, T.E. et al. (1991) *Cancer Res.* 51, 2077–2083.
7 Han, L. et al. (1990) *Blood* 75, 1234–1239.
8 Han, Z.C. et al. (1992) *J. Lab. Clin. Med.* 120, 645–660.
9 Tunnacliffe, A. et al. (1992) *Blood* 79, 2896–2900.
10 St. Charles, R. et al. (1989) *J. Biol. Chem.* 264, 2092–2099.
11 Mayo, K.H. and Chen, M.J. (1989) *Biochemistry* 28, 9469–9478.
12 Poncz, M. et al. (1987) *Blood* 69, 219–223.
13 Proudfoot, A.E.I. et al. (1995) *Eur. J. Biochem.* 228, 658–664.
14 Doi, T. et al. (1987) *Mol. Cell. Biol.* 7, 898–904.
15 Shigeta, O. et al. (1991) *Thromb. Res.* 64, 509–520.
16 Ramachandran, B. et al. (1995) *Exp. Hematol.* 23, 49–57.

NAP-2 | Neutrophil activating protein-2

Alternate names

Platelet basic protein (PBP) (parent molecule), connective tissue activating peptide III (CTAP-III) (precursor), low affinity platelet factor 4 (LA-PF4) or β-thromboglobulin (β-TG) (precursor), factor C

Family

α family (CXC family)

Molecule

NAP-2 is a proteolytic fragment of PBP corresponding to amino acids 25–94. The CTAP-III and LA-PF4 or β-TG, which are released from activated platelets, are inactive precursors of NAP-2. Leukocytes and leukocyte-derived proteases have been found to convert the inactive precursors into NAP-2 by proteolytic cleavage at the N-terminus[1].

Tissue sources

Platelets.

Target cells

Neutrophils, basophils, eosinophils, fibroblasts, NK cells, megakaryocytes, endothelial cells.

Physicochemical properties

Property	Human	Porcine
pI	9.0	8.7
Signal	1–34	?
Amino acids		
Precursor	128	119
Mature (PBP)	92	
Mature (CTAP-III)	84	
Mature (β-TG)	80	
Mature (NAP-2)	69	
Disulfide bonds	a.a. 63–89	
	a.a. 65–105	
Glycosylation sites		
Molecular weight		

Regulation of expression

NAP-2 expression is not transcriptionally regulated since it is produced by the cleavage of two inactive precursors CTAP-III and PBP, which are stored in the α-granules of blood platelets. CTAP-III is rapidly cleaved by chymotrypsin, cathepsin G and trypsin, yielding NAP-2. CTAP-III degradation by human neutrophil elastase and porcine pancreatic elastase does not yield NAP-2.

Conditioned medium from monocytes stimulated with LPS cleaves purified CTAP-III into NAP-2 through proteinases that were highly sensitive to PMSF, moderately sensitive to leupeptin, and insensitive to EDTA.

Coincubation of CTAP-III with PMN results in complete conversion of the precursor to NAP-2, as does incubation of CTAP-III with PMN-conditioned medium[2].

Release of β-TG from platelets can be significantly reduced by addition of prostaglandin PGE_1 and theophylline to whole blood and maintaining the blood at 4°C[3].

β-TG is released from platelets following stimulation with ADP, thrombin or collagen with time course and concentration dependence similar to that of serotonin release.

Expression in disease

Human studies

Plasma β-TG levels are significantly elevated in patients suffering from peripheral vascular disease but not those with cerebrovascular disease.

Both plasma and urine β-TG are significantly elevated in patients with deep vein thrombosis.

In patients in whom myocardial infarction had occurred more than 6 months previously, β-TG levels are significantly elevated. The β-TG level is unrelated to the extent of coronary artery disease. In patients with prior infarction, β-TG correlates directly with the extent of left ventricular regional dysfunction and inversely with ejection fraction.

Megakaryocytes

Inhibits megakaryocytopoiesis.

Neutrophils[4]

(CTAP-III is ineffective in inducing any of the neutrophil responses mentioned here.)
Calcium mobilization (100 nM)
Chemotaxis (0.3–10 nM)
Elastase release (1 nM)
Degranulation (100 nM)
Lactoferrin release (100 nM)
Respiratory burst
Exocytosis of β-glucosaminidase
Increase in the F-actin content

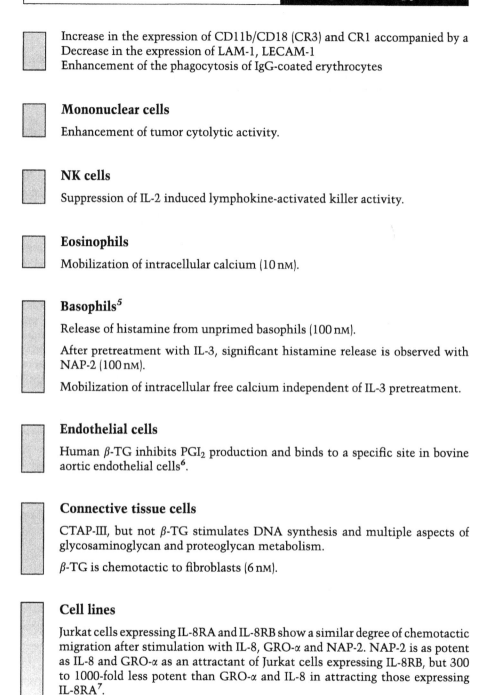

Increase in the expression of CD11b/CD18 (CR3) and CR1 accompanied by a Decrease in the expression of LAM-1, LECAM-1

Enhancement of the phagocytosis of IgG-coated erythrocytes

Mononuclear cells

Enhancement of tumor cytolytic activity.

NK cells

Suppression of IL-2 induced lymphokine-activated killer activity.

Eosinophils

Mobilization of intracellular calcium (10 nM).

Basophils[5]

Release of histamine from unprimed basophils (100 nM).

After pretreatment with IL-3, significant histamine release is observed with NAP-2 (100 nM).

Mobilization of intracellular free calcium independent of IL-3 pretreatment.

Endothelial cells

Human β-TG inhibits PGI$_2$ production and binds to a specific site in bovine aortic endothelial cells[6].

Connective tissue cells

CTAP-III, but not β-TG stimulates DNA synthesis and multiple aspects of glycosaminoglycan and proteoglycan metabolism.

β-TG is chemotactic to fibroblasts (6 nM).

Cell lines

Jurkat cells expressing IL-8RA and IL-8RB show a similar degree of chemotactic migration after stimulation with IL-8, GRO-α and NAP-2. NAP-2 is as potent as IL-8 and GRO-α as an attractant of Jurkat cells expressing IL-8RB, but 300 to 1000-fold less potent than GRO-α and IL-8 in attracting those expressing IL-8RA[7].

In Swiss 3T3 fibroblasts both CTAP-III and NAP-2, but not IL-8 increase glucose transport and the expression of both protein and mRNA levels of glucose transporter *GLUT1*.

Intracellular signaling

NAP-2 causes activation of GTPase by increasing V_{max} but not by decreasing K_m. High-affinity binding of GTP [^{35}S] to neutrophil plasma membranes is stimulated half-maximally and maximally (up to 5-fold) by NAP-2 at about 10 nM and 100 nM respectively

NAP-2 stimulation of neutrophils is independent of activation of PKC

Induces elevated levels of cAMP indicating the activation of adenylate-cyclase

Receptor binding characteristics

Competition of ^{125}I-labeled IL-8 with increasing concentrations of unlabeled NAP-2 resolves two classes of binding sites; about 70% of them bind NAP-2 with high affinity (K_d 0.34 nM), while 30% are of low affinity (K_d 100 nM).

The ^{125}I-labeled NAP-2 specifically binds to PMN with two different affinities ($K_d = 0.65$ and 22.4 nM). Unlabeled IL-8 completely competes NAP-2 binding. A total of 60 000–90 000 receptors per neutrophil are found for NAP-2[8]. Of these 30–45% are of high affinity with a mean K_d value of 0.3 nM for NAP-2, and 55–70% of low affinity ($K_d = 30$ nM) Two neutrophil membrane proteins of approximately 70 kD and 44 kD (p70 and p44) are specifically cross-linked with labeled NAP-2. Unlabeled IL-8 fully inhibits this cross-linking and the binding of labeled NAP-2 to the high-affinity sites on neutrophils or neutrophil membranes. Treatment of membranes with digitonin results in the preferential solubilization of a single receptor species, corresponding to p44, that binds NAP-2 with low affinity ($K_d = 30$ nM).

Exposure of neutrophil membranes to 100 μM GTPγS leads to a 75-fold increase of the K_d in approximately 60% of the receptors[8].

Binds erythrocytes at 1000–9000 sites/cell with a K_d of approximately 5 nM.

Does not compete with ^{125}I-N51 for binding to orphan receptor NIH-μIL-8Rβ.

Competes for binding to rabbit IL-8 receptor (5B1a) with an affinity binding profile of IL-8 >> NAP-2 > MGSA. The corresponding apparent K_i values for IL-8, NAP-2, and MGSA are 4, 120, and 320 nM, respectively.

Binds to COS-7 cells expressing IL-8 receptor clone (4ab) with low affinity.

Binds to Jurkat cells stably expressing IL-8RA with low affinity (K_d 200–500 nM) and those expressing IL-8RB with high affinity.

The gene ECRF3 of Herpesvirus saimiri virus which is a T-lymphotropic virus encodes a function receptor for NAP-2.

Human NAP-2 binds to rabbit neutrophil membranes with high affinity.

Mutants of NAP-2

The C-terminally truncated variant of NAP-2 exhibits 3-fold higher neutrophil-stimulating capacity and binding affinity than the full-size protein.

In the Ala-Glu-Leu-Arg (AELR) N-terminus of NAP-2, substitution of E or R abolishes Ca^{2+} mobilization and elastase secretion indicating that the N-terminus of NAP-2 is critical for high-level neutrophil-activating function. Synthetic NAP-2 (51–70) does not mobilize Ca^{2+}, indicating the C-terminus alone has no neutrophil activating properties.

Cross-desensitization

Short-term priming of PMN with a nonstimulatory dose of NAP-2 leads to drastic down-regulation of the subsequent degranulation response, by higher dosages of NAP-2, MGSA, or IL-8.

NAP-2, which binds almost exclusively to the IL-8RB, minimally desensitizes neutrophils to subsequent activation by IL-8 because IL-8 binds to and activates via both IL-8RA and B[9].

Calcium mobilization in basophils induced by NAP-2 is desensitized by pretreatment with IL-8.

NAP-2 pretreament of basophils cross-desensitizes their subsequent stimulation by IL-8 for the release of histamine and leukotriene.

NAP-2 pretreament of neutrophils desensitizes their subsequent stimulation by ENA-78.

Gene structure

The β-TG gene (*BTG1*) is 1139 bp long and, like other members of the small inducible gene (SIG) family, is divided into three exons. Southern blot analysis of genomic DNA suggests that, as with the *PF4* gene, there are multiple copies of the β-TG gene in the human genome. Two major transcriptional start sites are defined by primer extension analysis of platelet RNA, and, based on the more commonly used start site, the 5'-untranslated region is 87 bp. A TATA box is present beginning 32 bp upstream to this site. The first exon contains the 5'-untranslated region as well as the signal peptide. The second exon begins 6 bp 3' to the homologous site in *PF4*, and the third exon begins at a position homologous to that in *PF4* Interestingly, the β-TG and *PF4* genes have little detectable homology in the flanking or intronic sequences. In particular, a pyrimidine tract 5' to both the rat and human *PF4* genes is not present in the upstream region of the β-TG gene[10].

Gene location

β-TG genes (which are duplicate) are closely linked to the duplicated *PF4* genes and to other previously mapped CXC SIGs, namely *IL-8*, *GRO1*, *GRO2*, and *GRO3*, on a single 700 kb restriction fragment located in bands 4q12–q13[11].

By analysis of λ genomic clones, *BTG1* and *PF4* genes are separated by less than 7 kb, and the *BTG2* and *PF4*-alternate (*PF4V1*) genes by approximately 5 kb. Within each *BTG/PF4* duplication, the *BTG*-like gene is upstream of its linked *PF4*-like gene. The genes in this closely linked complex are expressed in a megakaryocyte-specific fashion.

Protein structure

Crystals of recombinant NAP-2 in which the single methionine at position 6 is replaced by leucine to facilitate expression belong to space group P1 (unit cell parameters $a = 40.8$, $b = 43.8$, and $c = 44.7$ Å and $\alpha = 98.4°$, $\beta = 120.3°$, and $\gamma = 92.8°$), with four molecules of NAP-2 ($M_r = 7600$) in the asymmetric unit. The molecular replacement solution, calculated with bovine PF4 as the starting model, is refined using rigid body refinement, manual fitting in solvent-leveled electron density maps, simulated annealing, and restrained least squares to an R-factor of 0.188 for 2σ data between 7.0 and 1.9 Å resolution. The final refined crystal structure includes 265 solvent molecules. The overall tertiary structure, which is similar to that of PF4 and IL-8, includes an extended N-terminal loop, three strands of antiparallel β-sheet arranged in a Greek key fold, and one α-helix at the C-terminus. The Glu-Leu-Arg sequence that is critical for receptor binding is fully defined by electron density and exhibits multiple conformations[12].

Amino acid sequence

Human PBP

```
1    MSLRLDTTPS CNSARPLHAL QVLLLLSLLL TALASSTKGQ TKRNLAKGKE
51   ESLDSDLYAE LRCMCIKTTS GIHPKNIQSL EVIGKGTHCN QVEVIATLKD
101  GRKICLDPDA PRIKKIVQKK LAGDESAD
```

CTAP-III, β-TG and NAP-2 represent proteolytic fragments of the mature PBP of 44–128, 48–128 and 59–128, respectively.

Porcine (NAP-2)

```
1    MSLRLGAISS CTTSSPFPVL QVLLPLSLLL TTLVPATMGA AKIEGRMAHV
51   ELRCLCLNTV SGIHPSNIQS LEVIRAGAHC AKVEVIATLK NDKKICLDPE
101  APRIKKIVQK IMEDGGSAA
```

Database accession numbers

	GenBank	SwissProt	PIR	MIM	Ref
Human CTAP-III	M54995 M11517	P02775	A24448 A37382 A39546	121010	10
Porcine	P43030				13

Despite overall similarity between the human and porcine proteins, the N-terminal region is almost completely different between the two species, with

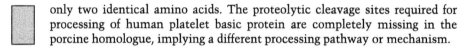
only two identical amino acids. The proteolytic cleavage sites required for processing of human platelet basic protein are completely missing in the porcine homologue, implying a different processing pathway or mechanism.

References

[1] Holt, J.C. et al. (1992) *Proc. Soc. Exp. Biol. Med.* 199, 171–177.

[2] Brandt, E. et al. (1991) *Cytokine* 3, 311–321.

[3] Ludlam, C.A. and Cash, J. D. (1976) *Br. J. Haematol.* 33, 239–247.

[4] Detmers, P.A. et al. (1991) *J. Immunol.*, 147, 4211–4217.

[5] Krieger, M. et al. (1992) *J. Immunol.* 149, 2662–2667.

[6] Hope, W. et al. (1979) *Nature* 282, 210–212.

[7] Loetscher, P. et al. (1994) *FEBS Lett.* 341, 187–192.

[8] Schnitzel, W. et al. (1991) *Biochem. Biophys. Res. Commun.* 180, 301–307.

[9] Moser, B. et al. (1991) *J. Biol. Chem.* 266, 10666–106671.

[10] Majumdar, S. et al. (1991) *J. Biol. Chem.* 266, 5785–5789.

[11] Wenger, R.H. et al. (1991) *Hum. Genet.* 87, 367–368.

[12] Malkowski, M.G. et al. (1995) *J. Biol. Chem.* 270, 7077–7087.

[13] Power, C.A. et al. (1994) *Eur. J. Biochem.* 221, 713–719.

Alternate names

CRG-2 or C7 gene product (mouse), mob-1

Family

α family (CXC family)

Molecule

IP-10 was originally identified as a gene product induced following stimulation of responding cells with IFN-γ. IP-10 lacks the ELR motif that is known to determine the biological significance of a given α-chemokine. The angiostatic potential of IP-10, which is opposite to the angiogenic effect of other ELR-containing α-chemokines, makes IP-10 a distinct member of the α-chemokine family.

Tissue sources

Endothelial cells, monocytes, fibroblasts, and keratinocytes[1]. High levels of IP-10 transcripts are found in lymphoid organs (spleen, thymus, and lymph nodes). Thymic and splenic stromal cells are found to express constitutively high levels of both IP-10 mRNA and protein, accounting for the high level of spontaneous expression in lymphoid tissue[2].

Target cells

Monocytes, progenitor cells, endothelial cells, NK cells

Physicochemical properties

Property	Human	Mouse
pI	~9	~9
Signal	1–21	1–21
Amino acids		
Precursor	98	98
Mature	77	77
Disulfide bonds	a.a. 30–57	30–57
	a.a. 32–74	32–74
Glycosylation sites	0	0
Molecular weight	10856	10789

Transcription factors

IFN-stimulated response element (ISRE) binding complex containing STAT1 or p91 and NF-κB1 (p50) and RelA (p65) are involved in synergistic activation of IP-10 by IFN-γ and TNF-α[2].

In the IP-10 promoter, a potential negative regulatory site for IFN-γ in the region between nucleotide positions −2002 and −930 and a positive regulator for LPS response in the region between bases -930 and -676 is reported. A 227-base fragment spanning positions -228 to -2 is the minimal sequence able to mediate LPS- and IFN-γ-dependent transcription of IP-10. Deletion of 24 bases, which include a highly conserved ISRE from the -228 construct, abolishes IFN-γ induced IP-10 expression. A 33-base fragment containing the IP-10 ISRE confers both IFN-γ and LPS sensitivity upon a heterologous promoter. Optimal response to IFN-γ requires both the ISRE and one of the two κB sites, whereas optimal response to LPS requires either both κB sites or one κB site and the ISRE. NF-κB binding complexes include NF-κB1 (p50) homodimers, and NF-κB1, RelA (p65) heterodimers of c-Rel.

Positive transcriptional response of IP-10 to IFN-γ requires an ISRE sequence motif and this site is the target of the suppressive action of IL-4 on IFN-γ-mediated transcription.

Transfection of macrophages with an expression vector encoding murine 1-κBα inhibits LPS-stimulated transcription of IP-10 driven by a 243 bp promoter sequence obtained from the 5' flanking region of the murine IP-10 gene.

The ISRE located between residues -204 and -228 is a primary target for the action of dsRNA on the IP-10 promoter.

Mob-1 is a downstream target of the Ras signaling pathway. The 417 bp mob-1 promoter, which contains dual NF-κB and AP-1 binding sites, confers Ras inducibility.

Regulation of expression

Stimulus	Cell type	Response
Cytokines		
IFN-γ	NIH 3T3 cells and murine macrophages	Synergistically enhances TNF-α and IL-2 induced levels of IP-10 mRNA
	Endothelial cells	IP-10 secretion
	Monocyte	IP-10 secretion
	Fibroblasts	IP-10 secretion
	Keratinocytes	IP-10 secretion
	Astrocytes and microglia	Induction of CRG-2 mRNA
α-, β-, and γ-IFNs	Murine macrophage cell line RAW 264.7	Induction of CRG-2 mRNA
Intradermal IL-2	Dermal inflammatory cells and keratinocytes of HIV-1-seropositive patients	Rapid up-regulation of IP-10
IL-1 or TNF-α[3]	Human synovial cells and fibroblasts	Expression of transcripts for IP-10
GM-CSF	Monocytes	A diminished induction of the gene for IP-10, in spite of increased IFN-γ receptor expression
IL-4	Macrophages	Inhibits IP-10 expression induced by LPS but not by IFN-γ/IL-2

Stimulus	Cell type	Response
	Monocytic leukemia cell line, THP-1	Inhibition of transcriptional activation of IP-10 by IFN-α and IFN-γ
Viruses		
Paramyxovirus, Newcastle disease virus	Astrocytes and microglia	Induce IP-10 expression within 2–6 h
Bacteria		
Mycobacterium tuberculosis	Murine macrophages	Induces rapid expression of IP-10 gene
Others		
LPS and lipid A	Murine peritoneal macrophages	Induce C7 mRNA expression
Phytohemagglutinin A (PHA) and phorbol 12-myristate 13-acetate	Human T-lymphocytes	Induce the expression of IP-10
Depletion of K^+ and Na^+	LPS-treated macrophages	Potentiation of IP-10 mRNA expression as a result of inhibition of Na^+/K^+ exchange through the Na^+K^+ ATPase
Double-stranded (ds) RNA	Fibroblasts and IFN-resistant Daudi cells	Induces IP-10 mRNA
Cyclosporin-A	T-lymphocytes	Blocks IP-10 expression induced with PHA/PMA
A synthetic flavone analog 5,6-dimethyl-xanthenone-4-acetic acid	Primary murine macrophages	Induces IP-10 expression
Taxol (1–30 mM)	Murine C3H/OuJ macrophages	Induction of IP-10 gene expression
Calmodulin antagonists such as trifluoperazine and Ca^{2+} antagonists such as bis-(o-amino-phenoxy)-ethane-$N,N,N'N'$-tetraacetic acid	Macrophages	Increase LPS-induced IP-10 mRNA levels
Okadaic acid and calyculin A, inhibitors of protein phosphatases 1 and 2A[4,5]	Murine peritoneal macrophages	Induce the expression of IP-10
Dexamethasone	Macrophages	Suppresses LPS-induction of IP-10 mRNA.
Sodium salicylate	Murine bone marrow stromal cells	Attenuates the IL-1 and LPS-induced enhancement of IP-10 mRNA
Oxidized LDL	Macrophages	Suppresses IP-10 mRNA expression induced by IFN-γ and IL-2 but not by LPS
Oncogenic Ras and serum growth factors that activate endogenous Ras	NIH 3T3 cells	Induce mob-1 expression but oncogenic induction is constitutive whereas the serum induction is transient

Expression in disease

Human studies

Human delayed-type response to purified protein derivative of tuberculin (PPD) elicits the acute expression of IP-10 in dermal macrophages and endothelial cells, the basal layer of epidermal keratinocytes and chronic expression in entire epidermis. Keratinocytes are the major producers of IP-10[6].

Human rIL-2 (10–30 μg) injection intradermally into the skin of patients with lepromatous leprosy with high bacillary loads leads to the expression of IP-10 by keratinocytes.

In cutaneous T-cell lymphoma lesions high expression of IP-10 extending from the epidermal to the suprabasal keratinocytes is seen[7].

IP-10 protein and mRNA are detected in keratinocytes and the dermal infiltrate from active psoriatic plaques. Successful treatment of active plaques decreases IP-10 expression in plaques.

Expression of IP-10 mRNA is seen in alveolar macrophages of HIV-seropositive individuals but not normal controls.

Animal studies

Expression of IP-10 mRNA in the glomeruli is seen in a murine model of anti-glomerular basement membrane antibody induced glomerulonephritis.

Astrocytes express IP-10 mRNA in a mouse model of experimental autoimmune encephalomyelitis[8]. IP-10 expression was seen in the spinal cord 1–2 days before clinical signs were apparent.

IP-10 mRNA is expressed in the fibroblast-like cells in the skin of mice during contact hypersensitivity reactions to the hapten trinitrochlorobenzene (TNCB).

Peak expression of the IP-10 gene is seen at 3 h in heterologous-phase anti-GBM antibody-induced glomerulonephritis in Lewis rats.

Intravenous administration of IFN-γ and TNF-α to mice induces the expression of IP-10 mRNA in the adherent cell population of the spleen, and resident peritoneal macrophages in a rapid and transient manner in both liver and kidneys.

Inhaled IFN-γ induces the expression of IP-10 in bronchoalveolar fluid macrophages.

Elevated expression of the IP-10 gene is found in murine renal cell carcinoma from mice treated with a combination of IFN-α and IL-2 for 4 days.

In mice exposed to hypoxia (PO$_2$ approximately 30–40 Torr), increased pulmonary leukostasis and increased myeloperoxidase activity in tissue homogenates is paralleled by increased levels of transcripts for IP-10 in hypoxic lung tissue.

When expressed in tumor cells IP-10 has no effect on the growth of the tumor cells in culture, but elicits a powerful host-mediated antitumor effect that is

T lymphocyte dependent, noncell autonomous, and mediated by the recruitment of an inflammatory infiltrate composed of lymphocytes, neutrophils, and monocytes.

IP-10 is a potent inhibitor of both IL-8 and βFGF-induced angiogenic activity in both *in vitro* and *in vivo* assays of angiogenesis[9].

The intracerebroventricular administration of IP-10 decreases the short-term (2 h) food intake in rats.

In vitro biological effects

Progenitor cells

IP-10 administration causes suppression of colony formation of granulocyte-macrophage, erythroid, and multipotential progenitor cells and early human bone marrow progenitor cells which need steel factor and GM-CSF or SLF and erythropoeitin.

Monocytes

Chemotaxis

T lymphocytes

Chemotaxis of IL-2 stimulated CD4+ and CD29+ T lymphocytes
Enhanced ability to bind to an IL-1-treated endothelial cell monolayer

NK cells[10]

Chemotaxis
Release of granule-derived serine esterases
Augmentation of NK-specific cytolytic responses

Endothelial cells

Suppresses endothelial cell differentiation into tubular capillary structures
IP-10 inhibits endothelial cell proliferation ($IC_{50} = 150$ nM).

Cell lines

Suppression of human hematopoietic cell line, MO7e cells.

Pretreatement of MO7e cells with IP-10 increases in intracellular cAMP levels, and Raf-1 phosphorylation. IP-10 antagonizes the synergistic action of GM-CSF and SLF by inactivation of Raf-1 and the down-regulation of protein synthesis.

Receptor binding characteristics

IP-10 binds to endothelial, epithelial, and hematopoietic cells with a K_d of 25 nM and this binding is dependent on the presence of cell surface heparan sulfate proteoglycans (HSPG). PF4, but not IL-8, MCP-1, RANTES, MIP-1α, or MIP-1β, can compete effectively with IP-10 for binding to the cell surface[11].

Gene structure

Mouse

The genomic organization of IP-10 reveals three introns that interrupt the transcribed sequence into four functional domains. IP-10 and PF4 genes are interrupted in precisely the same positions within homologous codons. A DNase I-hypersensitive site is found in responsive cells in a region upstream of the transcription initiation site. This hypersensitive site is induced by IFN-γ and thus provides a structural basis for the transcriptional activation seen for this gene by IFN-γ.

Gene location

IP-10 is located on chromosome 4 at band q2, proximal to a locus associated with an acute monocytic/B-lymphocyte lineage leukemia that exhibits the nonrandom translocation (4;11) (q21;q23). Restriction fragment length polymorphism in the 5' region of the IP-10 gene is observed.

Amino acid sequence

Human

```
 1  MNQTAILICC LIFLTLSGIQ GVPLSRTVRC TCISISNQPV NPRSLEKLEI
51  IPASQFCPRV EIIATMKKKG EKRCLNPESK AIKNLLKAVS KEMSKRSP
```

Mouse

```
 1  MNPSAAVIFC LILLGLSGTQ GIPLARTVRC NCIHIDDGPV RMRAIGKLEI
51  IPASLSCPRV EIIATMKKND EQRCLNPESK TIKNLMKAFS QKRSKRAP
```

Database accession numbers

	GenBank	SwissProt	PIR	MIM	Ref
Human	X02530	P02778	A03243	147310	6
Mouse	M86829	P17515	A35498		12
	M33266		A35812		
	M37136		A45492		

References
[1] Luster, A.D. and Ravetch, J.V. (1987) *J. Exp. Med.* 166, 1084–1097.
[2] Gattass, C.R. et al. (1994) *J. Exp. Med.* 179, 1373–1378.

3 Bedard, P.A. and Golds, E.E. (1993) *J. Cell Physiol.* 154, 433–441.

4 Tebo, J.M. and Hamilton, T.A. (1994) *Cell Immunol.* 153, 479–491.

5 Luong, H. et al. (1994) *Biochem. J.* 299, 799–803.

6 Ohmori, Y. and Hamilton, T.A. (1995) *J. Immunol.* 154, 5235–5244.

7 Kaplan, G. et al. (1987) *J. Exp. Med.* 166, 1098–1108.

8 Ransohoff, R.M. et al. (1993) *FASEB J.* 7, 592–600.

9 Strieter, R.M. et al. (1995) *Biochem. Biophys. Res. Commun.* 210, 51–57.

10 Taub, D. D. et al. (1995) *J. Immunol.* 155, 3877–3888.

11 Luster, A.D. et al. (1995) *J. Exp. Med.* 182, 219–231.

12 Ohmori, Y. and Hamilton, T.A. (1990) *Biochem. Biophys. Res. Commun.* 168, 1261–1267.

MIP-2 | Macrophage inflammatory protein-2

Alternate names

MIP-2α (or GRO-β gene product), MIP-2β (or GRO-γ gene product), leukocyte-derived neutrophil chemotactic factor 2 (LDNCF-2), cytokine induced neutrophil chemoattractant 3 (CINC-3). Two MIP-2 related proteins (CINC-2α and CINC-2β) are also reported.

Family

α family (CXC family)

Molecule

The MIP-2 class consists of MIP-2α, also known as the GRO-β gene product, and MIP-2β or the GRO-γ gene product. The role of MIP-2 is best characterized in anti-GBM antibody-induced glomerulonephritis in mice.

Tissue sources

Mast cells, cardiac myocytes, mesangial cells, alveolar macrophages, epidermal cells, human nasal and bronchial epithelium.

Target cells

Neutrophils[1], basophils, epithelial cells.

Physicochemical properties

Property	Human (MIP-2α)	Human (MIP-2β)	Mouse	Rat
pI	~9.7	~9.7	?	?
Signal	1–34	1–34	1–27	1–27
Amino acids				
Precursor	107	107	100	100
Mature	73	73	73	73
Disulfide bonds	a.a. 43–69	a.a. 43-69	a.a. 36-62	a.a. 36–62
	a.a. 45–85	a.a. 45–85	a.a. 38–78	a.a. 38–78
Glycosylation sites	0	0	0	0
Molecular weight	11 389	11 345	10 621	10 783

Transcription factors

Human MIP-2α promoter contains an LPS-responsive region between residues −104 and +30. This region contains two κB sequence motifs. The first motif (position −70 to −59, κB1) is highly conserved and is essential for LPS-inducibility of MIP-2α[2].

Murine MIP-2 promoter contains an LPS-responsive element in the region that contains a conserved NF-κB consensus motif and lies 51 to 70 bp 5' from the transcription start site.

Regulation of expression

Rat peritoneal connective tissue-type mast cells stimulated with anti-IgE for 4 h show elevated expression of mRNA encoding MIP-2.

Cultured myocytes stimulated with IL-1, TNF, or LPS express MIP-2α mRNA.

Mesangial cells express MIP-2 mRNA in response to mediators of acute renal inflammation such as IL-1β, TNF-α, and LPS, but not chronic renal inflammation such as TGF-β[3].

Dexamethasone suppresses ozone-induced MIP-2 mRNA expression and neutrophil accumulation in the lung. Dexamethasone also suppresses MIP-2 expression in trachea homogenates from rats exposed to SO_2.

Expression in disease

Human studies

The majority of synovial fibroblast cell lines derived from osteoarthritic or non-inflammatory synovia show a relative increase in the constitutive expression of GRO-β when compared to synovial fibroblasts obtained from rheumatoid synovia[4].

Animal studies

Expression of MIP-2 mRNA is seen in the glomeruli in a murine model of anti-GBM antibody-induced glomerulonephritis[5]. Injection of anti-MIP-2 antibody 30 min before anti-GBM antibody is effective in reducing neutrophil influx and periodic acid–Schiff deposits containing fibrin[6].

Mice infected intracerebrally with *Listeria monocytogenes* express MIP-2 in the brain in addition to the β-chemokines, MIP-1α and MIP-1β. The cellular sources of these chemokines comprise both the blood-derived PMNs and monocytes infiltrating the meninges, the ventricular system, and the periventricular area.

Following ozone exposure of rats, MIP-2 mRNA expression is seen in the lung which peaks at 2 h after exposure and slowly declines thereafter.

MIP-2 is the primary chemotactic factor for neutrophil infiltration in a rat allergic inflammation model induced by injection of antigen solution into the air pouch.

In a rat model of SO_2-induced bronchitis, trachea homogenates contain MIP-2 mRNA after 1 day of SO_2 exposure.

Inhibition of IL-10 bioactivity *in vivo* in a mouse model of LPS-induced endotoxemia results in a greater and more sustained increase in plasma MIP-2 levels and is accompanied by early increases in lung polymorphonuclear leukocyte influx and lung capillary leak. Anti-IL-10 mediated lethality is abrogated by concomitant treatment with anti-MIP-2 antibody[7].

In a mouse model of collagen-induced arthritis, detectable levels of MIP-2 are first observed between days 32 and 36, after initial type II collagen challenge.

Antibodies against MIP-2 delay the onset of arthritis and reduce the severity of arthritis. Anti-IL-10 treatment increases the expression of MIP-2 and exacerbates the disease.

Enhancement of MIP-2 in lung homogenate is observed following the challenge of CD-1 mice by *Klebsiella pneumoniae* intratracheally.

Transgenic rats carrying the human T-lymphotropic virus type I pX gene (pX rats) develop undifferentiated mammary carcinomas expressing high levels of MIP-2 and are characterized by massive infiltration of granulocytes in the tumor tissue[8].

Elevated expression of MIP-2 mRNA is seen in uvea and retina during endotoxin-induced uveitis in the rat.

A rapid increase in MIP-2 mRNA levels is seen in bronchoalveolar lavage cells and whole trachea homogenates after LPS instillation intratracheally in rats.

A marked induction of MIP-2 is seen *in vivo* in alveolar macrophages ingesting albumin-coated fluorescent latex particles opsonized with anti-albumin IgG.

In a mouse model of accelerated rejection of cardiac allografts, a maximal level of MIP-2 mRNA is seen at 3–6 h after transplantation. Treatment with rapamycin completely abrogates MIP-2 induction in cardiac tissue.

MIP-2 is one of the two major chemoattractants for neutrophils isolated from rat inflammatory exudate induced by a subcutaneous injection of LPS[9].

Elevated levels of MIP-2 mRNA are detected on day 3 in a murine model of wound healing.

Cultured liver slices obtained from acute or subchronic cadmium-exposed rats and mice express mRNA transcripts of MIP-2.

In vitro biological effects

Progenitor cells

MIP-2β possesses direct suppressive activity on single stem/progenitor cells from adult human bone marrow in the presence or absence of serum[10].

MIP-2α suppresses the colony formation of immature subsets of myeloid progenitor cells stimulated by GM-CSF plus steel factor.

Neutrophils

GRO-β is chemotactic rather than chemokinetic whereas GRO-γ is chemokinetic as well as chemotactic[1].
Calcium mobilization[11]
Chemotaxis (2 ng/ml)[11]
Exocytosis of elastase[11]
Respiratory burst[11]
Loss of L-selectin and an increase in MAC-1 expression
Rat MIP-2 is chemotactic to human neutrophils.

CINC-3 induces an increase in intracellular $[Ca^{2+}]$ and release of cathepsin G in rat neutrophils.

Epithelial cells

Rat MIP-2 is mitogenic.

Basophils

Chemotaxis
Intracellular calcium concentration changes

Receptor binding characteristics

Human MIP-2 binds specific receptors on human blood neutrophils with high affinity (2–4 nM). Both MIP-2α and MIP-2β compete for up to 60% of the IL-8 binding sites on neutrophils whereas IL-8 competes for almost 100% of either of the MIP-2 binding sites[12].

MIP-2α and MIP-2β compete for ^{125}I-KC binding to mouse neutrophils with high affinity.

MIP-2 binds with high affinity (4 nM) to COS cells transiently expressing murine IL-8RB receptor homologue (muIL-8RB). MIP-2 competes with ^{125}I-N51 for binding to NIH 3T3 cells constitutively expressing muIL-8RB.

Cross-desensitization

Rat GRO-α does not cross-desensitize CINC-3 in the induction of an increase in intracellular $[Ca^{2+}]$ in rat neutrophils.

Gene structure

The murine MIP-2 genomic clone displays the canonical four exon/three intron structure typical of other genes in the chemokine α family. Potential *cis* regulatory elements in the proximal promoter region are highly conserved between muMIP-2 and its three most closely related human homologs: human GRO-α, GRO-β, and GRO-γ.

Gene location

The human MIP-2 gene is located on chromosome 4 on the proximal long arm at 4q21.

Protein structure

No structural information is reported. Based on the amino acid sequence similarity to MGSA, the MIP-2 tertiary structure is thought to be similar to MGSA.

Amino acid sequence

Human MIP-2α (GRO-β)

```
1   MARATLSAAP SNPRLLRVAL LLLLLVAASR EAAGAPLATE LRCQCLQTLQ
51  GIHLKNIQSV KVKSPGPHCA QTEVIATLKN GQKACLNPAS PMVKKIIEKM
101 LKNGKSN
```

Human MIP-2β (GRO-γ)

```
1   MAHATLSAAP SNPRLLRVAL LLLLLVAASR RAAGASVVTE LRCQCLQTLQ
51  GIHLKNIQSV NVRSPGPHCA QTEVIATLKN GKKACLNPAS PMVQKIIEKI
101 LNKGSTN
```

Mouse (MIP-2)

```
1   MAPPTCRLLS AALVLLLLLA TNHQATGAVV ASELRCQCLK TLPRVDFKNI
51  QSLSVTPPGP HCAQTEVIAT LKGGQKVCLD PEAPLVQKII QKILNKGKAN
```

Rat (MIP-2)

```
1   MAPPTRQLLN AVLVLLLLLA TNHQGTGVVV ASELRCQCLT TLPRVDFKNI
51  QSLTVTPPGP HCAQTEVIAT LKDGHEVCLN PEAPLVQRIV QKILNKGKAN
```

Rat CINC-2β is a 68-residue chemoattractant produced by cleavage of a 32-residue signal peptide. The difference in amino acid sequences between CINC-2α and CINC-2β consisted of only three C-terminal residues.

Database accession numbers

	GenBank	SwissProt	PIR	MIM	Ref
MIP-2α					
Human	X53799	P19875	JH0281	139110	13
	M36820				
	M57731				
Mouse	X53798	P10889	JH0200		14
Rat	X65647	P30348			15
MIP-2β					
Human	X53800	P19876	B38290	139111	13
			JH0282		
	M36821				
	G183633				

References

1 Zagorski, J. et al. (1994) *Protein Expr. Purif.* 5, 337–345.
2 Ohmori, Y. et al. (1995) *J. Immunol.* 155, 3593–3600.
3 Wu, G.J. et al. (1995) *Am. J. Physiol.* 269, F248–256.
4 Hogan, M. et al. (1994) *Cytokine* 6, 61–69.
5 Tang, W.W. et al. (1995) *Am. J. Physiol.* 269, F323–330.
6 Feng, L. et al. (1995) *J. Clin. Invest.* 95, 1009–1017.
7 Standiford, T.J. et al. (1995) *J. Immunol.* 155, 2222–2229.

[8] Yamada, S. et al. (1995) *Cancer Res.* 55, 2524–2527.

[9] Watanabe, K. et al. (1993) *Eur. J. Biochem.* 214, 267–270.

[10] Lu, L. et al. (1993) *Exp. Hematol.* 21, 1442–1446.

[11] Geiser, T. et al. (1993) *J. Biol. Chem.* 268, 15419–15424.

[12] Suzuki, H. et al. (1994) *J. Biol. Chem.* 269, 18263–18266.

[13] Tekamp-Olson, P. et al. (1990) *J. Exp. Med.* 172, 911–919.

[14] Wolpe. S.D. et al. (1989) *Proc. Natl. Acad. Sci. USA* 86, 612–616.

[15] Driscoll, K. (1992) Direct GenBank submission.

ENA-78
Epithelial derived neutrophil attractant-78

Family

α family (CXC family)

Molecule

ENA-78 is related to NAP-2 and GRO-α (sequence identity 53% and 52%, respectively) and IL-8 (22% identity).

Tissue sources

Epithelial cells, platelets[1].

Target cells

Neutrophils.

Physicochemical properties

Property	Human	Bovine
pI	8.8	~9
Signal	1 – ?	1 – ?
Amino acids		
Precursor	114	48 (partial)
Mature	114	
Disulfide bonds	a.a. 49–75	
	a.a. 51–91	
Glycosylation sites		
Molecular weight		
Precursor	11 972	
Mature	8357	

Transcription factors

The transcription start site of the ENA-78 gene is mapped to a position 96 bp upstream from the translation initiation site. The expression of the ENA-78 gene is induced by TNF-α, IL-1β, or PMA. The 125 bp promoter region contains binding sites for C/EBP and NF-κB. Transfection of 293 cells with promoter deletion mutants demonstrates that the NF-κB element, but not the C/EBP site, is sufficient for expression and induction by either TNF-α or IL-1β. In contrast, the IL-8 gene requires both elements[2]. The 5' flanking region contains potential binding sites for several nuclear factors such as AP-2 and IRF-1[3].

Regulation of expression

IL-1β and TNF-α stimulation induce the expression of ENA-78 in the human type II epithelial cell line A549.

IL-1β stimulation of renal cortical epithelial cells with tubular cell attributes leads to increased steady state levels of ENA-78 mRNA and increased secretion. However, TNF-α, IFN-γ, and LPS do not stimulate ENA-78 steady state mRNA or antigenic peptide production by renal cortical epithelial cells[4]. IFN-α and IFN-γ down-regulate the production of ENA-78 in LPS- and IL-1-stimulated human monocytes.

Expression in disease
Human studies

ENA-78 is found at elevated levels in peripheral blood, synovial fluid, and synovial tissue from RA patients. Anti-ENA-78 antibodies neutralize 42% of the chemotactic activity for PMNs found in RA synovial fluid. Isolated RA synovial tissue fibroblasts *in vitro* constitutively produce significant levels of ENA-78. In addition, RA and osteoarthritis synovial tissue fibroblasts as well as RA synovial tissue macrophages constitutively produce ENA-78. ENA-78 is predominantly found in synovial lining cells, followed by macrophages, endothelial cells, and fibroblasts from the synovial tissue of RA patients[5].

Biopsy tissue from acutely rejecting human renal allografts has higher levels of ENA-78 mRNA compared with nonrejecting renal allograft controls.

Animal studies

The presence of bovine ENA in the hyperplastic type II alveolar epithelial cells and in pulmonary alveolar leukocytes of pneumonic bovine lungs strongly supports a role for ENA-78 in the genesis of pulmonary inflammation.

In the context of hepatic ischemia/reperfusion injury, production of pulmonary-derived ENA-78 correlates with an increase in pulmonary microvascular permeability and lung neutrophil sequestration. Passive immunization of the animals with neutralizing TNF antiserum results in a significant suppression of pulmonary-derived ENA-78 and passive immunization with neutralizing ENA-78 antiserum results in an attenuation of pulmonary neutrophil sequestration and microvascular permeability[6].

In vitro biological effects
Neutrophils

ENA-78 stimulates neutrophils, inducing chemotaxis, a rise in intracellular free calcium and exocytosis.

Cross-desensitization

ENA-78 cross-desensitizes human neutrophils for further stimulation with IL-8, NAP-2, and GRO-α.

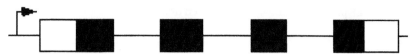

Gene structure

The open reading frame of 342 nucleotides in the *ENA-78* gene encodes for a 114 amino acid protein.

Gene location

The human *ENA-78* gene has been mapped to chromosome 4q13–q21, the same locus as several other inflammatory chemokine genes.

Amino acid sequence

Human

```
  1  MSLLSSRAAR VPGPSSSLCA LLVLLLLLTQ PGPIASAGPA AAVLRELRCV
 51  CLQTTQGVHP KMISNLQVFA IGPQCSKVEV VASLKNGKEI CLDPEAPFLK
101  KVIQKILDGG NKEN
```

Bovine (partial amino acid sequence)

```
  1  VVRELRCVCL TTTPGIHPKT VSDLQVIAAG PVCSKVEVIAT LKNGXXV
```

Bovine ENA shows structural (73% identity in amino acid sequence) and functional homology to human ENA-78

Database accession numbers

	GenBank	SwissProt	PIR	MIM	Ref
Human	X78686	P42830		600324	1
	S69618				
	L37036				
	U12709				

References
[1] Power, C.A. et al. (1994) *Gene* 151, 333–334.
[2] Chang, M.S. et al. (1994) *J. Biol. Chem.* 269, 25277–25282.
[3] Corbett, M.S. et al. (1994) *Biochem. Biophys. Res. Commun.* 205, 612–617.
[4] Schmouder, R.L. et al. (1995) *Transplantation* 59, 118–124.
[5] Koch, A.E. et al. (1994) *J. Clin. Invest.* 94, 1012–1018.
[6] Colletti, L.M. et al. (1995) *J. Clin. Invest.* 95, 134–141.

EMF-1   Embryo fibroblast protein-1

Alternate names

pCEF precursor, CEF-4, 9E3, chicken or avian GRO

Family

α family (CXC family)

Molecule

EMF-1 has been reported in chicken fibroblasts and mononuclear cells and the occurrence of a human or murine homolog is not known. The avian gene 9E3/CEF-4, is expressed abundantly in exponentially growing cultures of chick embryo fibroblasts (CEFs) and at high levels in CEFs transformed with RSV[1]. The gene for EMF-1 was isolated from RSV-transformed CEF identified by differential screening of a cDNA library. EMF-1 is included in the chemokine family because of its sequence similarity to CTAP-III and PF4. EMF-1 mRNA is expressed at the G0–G1 transition and during the first G1 phase of normal CEFs reentering the cell cycle[2].

Tissue sources

RSV-infected cells

Target cells

Fibroblasts, mononuclear cells

Physicochemical properties

Property	Avian
Signal	1–16 (potential)
Amino acids	
Precursor	103
Mature	87
Disulfide bonds	a.a. 33–60
	a.a. 35–76
Glycosylation sites	0
Molecular weight	
Precursor	11 056

Transcription factors

Studies to identify *cis*-acting transcriptional elements that confer inducibility by v-*src* resulted in the identification of a region 1.53 kb upstream of the transcription start site, that confers a small degree of inducibility by v-*src*. Two potential AP-1 sites are present in the EMF-1 promoter. These elements alone do not confer a significant inducibility by v-*src* in primary CEFs[3].

Three sequences, two corresponding to binding sites for AP-1, PRD II/κB, and TAACGCAATT, define the src-responsive unit (SRU) of the EMF-1/CEF-4 promoter. In constructs containing a deletion of the SRU, multiple copies of AP-1 or PRD II/κB, but not TAACGCAATT, lead to activation of the promoter. Protein binding to AP-1, PRD II/κB, and TAACGCAATT are more abundant in the nuclei of transformed cells. No significant increase in EMF-1/CEF-4 promoter activity is detected early after activation of pp60 v-src. In contrast, a substantial activation of the EMF-1/CEF-4 promoter is detected late after a temperature shift. Factors interacting with the TAACGCAATT, PRD II/κB, and AP-1 elements accumulate gradually over a period of several hours[3].

Regulation of expression

Animal studies

The chicken EMF-1 gene is expressed abundantly in the cells of proliferating cultures but at very low levels in confluent cultures. pp60 v-src and serum are the only reported inducers of EMF-1[4].

EMF-1 mRNA expression is 20-fold higher in CEFs following RSV transformation because of increased transcription. The expression of EMF-1 mRNA in RSV-transformed CEF appears to be the result of transcriptional activation and mRNA stabilization.

When CEFs are wounded by scraping swaths across confluent cultures, EMF-1 is (1) transiently expressed after 'wounding' or serum-stimulation, (2) expressed in a cell cycle phase-dependent manner; it is triggered during the G0–G1 transition or early in G1 and subsides during S phase, and (3) stimulated to high levels by αFGF, βFGF, TGF-α, and TGF-β, to intermediate levels by PDGF and not stimulated by EGF. Cells that are constantly cycling do not express EMF-1, indicating that they either skip the portion of the cell cycle where 9E3 is induced or that they constitutively express a repressor of transcription or an RNA-degrading enzyme[5].

When the 3' untranslated region of the reporter gene is replaced with the EMF-1 3' untranslated region, the resulting construct is strongly responsive to stimulation by v-src. In addition, the EMF-1 3' untranslated region increases the response to serum and the tumor promoter PMA. This suggests that a post-transcriptional mechanism plays a major role in the induction of EMF-1 expression.

EMF-1/CEF-4 mRNA is expressed in chicken peripheral blood monocytes and its expression is stimulated by incubation of the monocytes with LPS or PMA.

In vivo studies

Animal studies

In situ hybridization and RNA blot analysis of the pattern of EMF-1 mRNA distribution indicates high expression in specific tissues in normal wings;

whereas connective tissue, tendon, and bone express high levels of the gene, muscle fibers, endothelium, epidermis, and bone marrow do not. The distribution coincides with that of interstitial collagen. Wounding results in marked elevation of the mRNA within the granulation tissue formed during healing and in adjacent tissues, especially those showing neovascularization. Similar elevation of mRNA occurs immediately adjacent to RSV tumors but, surprisingly, the tumor tissue itself shows no detectable levels of this message. Cells explanted from the tumors and grown in culture also show no expression of EMF-1, in marked contrast to the very high level found in similarly cultured RSV-transformed CEFs. These results show that there are intrinsic differences between transformed embryonic cells in tissue culture and RSV target cells in the hatched chick. The abundant expression of EMF-1 in normal tissues indicates that the product of this gene plays a normal physiological role in tissues growing by cell division, perhaps as a growth regulator. The elevated expression of EMF1 in areas of neo-vascularization, makes it possible that the product of this gene could act as an angiogenic factor. EMF-1 expression in conjunction with high collagen levels and in wounded tissues may point to a role in wound response and/or repair[1].

In newly hatched chicks, immunostaining of EMF-1 using a polyclonal antibody shows an abundant expression of the protein in the cells and extra-cellular matrix (ECM) of connective tissue and other tissues of mesenchymal origin, such as bone and tendon. Most cells in the granulation tissue of wounds show expression of EMF-1 protein, some more intensely than others; the EMF-1 expression is abundant in ECM, especially in areas of scar tissue where collagen is abundant. In RSV-induced tumors, the protein is absent except in necrotic areas where a few cells – potentially macrophages – express EMF-1[2].

In vitro biological effects

Mononuclear cells

EMF-1 is chemotactic for chicken peripheral blood mononuclear cells, as well as for heterophils.

Fibroblasts

Untransformed CEF and CEF transformed with RSV migrate in response to EMF-1.

EMF-1 is weakly mitogenic for CEF, causing a doubling of [^{3}H]-thymidine uptake when added to serum-starved CEF. EMF-1 is found to be associated not only with the cell and in the culture medium of RSV-transformed CEF but also with the extracellular matrix. The *in vivo* role of EMF-1 may be involved with chemotaxis and metastasis, rather than with direct stimulation of mitogenicity.

Amino acid sequence

Chicken

```
1   MNGKLGAVLA LLLVSAALSQ GRTLVKMGNE LRCQCISTHS KFIHPKSIQD
51  VKLTPSGPHC KNVEIIATLK DGREVCLDPT APWVQLIVKA LMAKAQLNSD
101 APL
```

Database accession numbers

	GenBank	SwissProt	PIR	Ref
Chicken	M16199 J02975	P08317	A26736	6

References
[1] Martins-Green, M. and Bissell, M. (1990) *J. Cell Biol.* 110, 581–595.
[2] Bedard, P.A. et al. (1987) *Proc. Natl. Acad. Sci. USA* 84, 6715–6719.
[3] Blobel, G.A. and Hanafusa, H. (1991) *Proc. Natl. Acad. Sci. USA* 88, 1162–1166.
[4] Sugano, S. et al. (1991) *Cell Regul.* 2, 739–752.
[5] Martins-Green, M. et al. (1992) *J. Cell Sci.* 101, 701–707.
[6] Sugano, S. et al. (1987) *Cell* 49, 321–328.

GCP-2 | Granulocyte chemotactic protein-2

Family

α family (CXC family)

Molecule

Human GCP-2 is found to be coproduced with IL-8 by osteosarcoma cells. The bovine homolog of human GCP-2 is also found in kidney tumor cells. Both human and bovine GCP-2 are chemotactic to human granulocytes and activate postreceptor mechanisms leading to release of gelatinase B which is indicative of a possible role in inflammation and tumor cell invasion.
Human and bovine GCP-2 are 67% similar at the amino acid level.

Tissue sources

Osteosarcoma cells, kidney tumor cells[1]

Target cells

Granulocytes

Physicochemical properties

Property	Human
Amino acids	
Mature	75
Variant	73 (missing a.a. 1–2 in the N-terminus)
Variant	70 (missing a.a. 1–5 in the N-terminus)
Variant	67 (missing a.a. 1–8 in the N-terminus)
Disulfide bonds	a.a. 12–38
	a.a. 14–54
Glycosylation sites	0
Molecular weight	7534

In vivo studies

Following intradermal injection of 200 ng/site, GCP-2 provokes a significant granulocyte infiltration, albeit to a lesser extent than do IL-8 and GRO-α. GCP-2 does not attract monocytes *in vivo*[2].

In vitro biological effects

Neutrophils[2]

Chemotaxis (3–10 nM)
Gelatinase B release (5 nM)

Protein structure

Both human and bovine GCP-2 occur in at least four N-terminally truncated forms. These 5–6 kD proteins do not differ in potency and efficacy as granulocyte chemotactic factors using a standard *in vitro* migration assay.

Amino acid sequence

Human

```
1   GPVSAVLTEL RCTCLRVTLR VNPKTIGKLQ VFPAGPQCSK VEVVASLKNG
51  KQVCLDPEAP FLKKVIQKIL DSGNK
```

Bovine

```
1   GPVAAVVREL RCVCLTTTPG IHPKTVSDLQ VIAAGPQCSK VEVIATLKNG
51  REVCLDPEAP LIKKIVQKIL DSGKN
```

Database accession numbers

	GenBank/EMBL	SwissProt	PIR	Ref
Human	P80162	A54188	138965	1
Bovine	P80221	B54188		2

References
[1] Proost, P. et al. (1993) *J. Immunol.* 150, 1000–1010.
[2] Proost, P. et al. (1993) *Biochemistry* 32, 10170–10177.

Stromal cell derived factor-1

Alternate names

Pre-B cell growth stimulating factor precursor (PBSF)

Family

α family (CXC family)

Molecule

SDF-1 was found to be produced by a stromal cell line, PA6, distinct from IL-7 and stem cell factor and supported the proliferation of a stromal cell-dependent pre-B-cell clone, DW34. Alternative splicing of the *SDF-1* gene results in two similar products, *SDF-1α* and *SDF-1β*. Strong evolutionary conservation and unique chromosomal localization of the *SDF-1* gene suggest that SDF-1 may have important functions distinct from those of other members of the intercrine family. SDF-1 appears to be the natural ligand for the LESTR receptor.

Tissue sources[1]

Stromal cells, bone marrow, liver tissue, and muscle.

Target cells

Pre-B cells[1].

Physicochemical properties

Property	Human	Mouse
Signal	1–18	1–18
Amino acids		
Precursor	89 (SDF-1α)	89
Precursor	93 (SDF-1β)	
Mature	71 (SDF-1α)	71
Mature	75 (SDF-1β)	
Disulfide bonds	a.a. 30–55	a.a. 30–55
	a.a. 32–71	a.a. 32–71
Glycosylation sites	0	0
Molecular weight		
Precursor	10 032	10 032

In vitro biological effects

T lymphocytes

Transient elevation of $[Ca^{2+}]_i$ in human tumor-infiltrating T lymphocytes and in cultured, IL-2-activated human peripheral blood-derived lymphocytes.

Chemotaxis.

Gene structure

The genomic structure of the *SDF-1* gene reveals that human SDF-1α and SDF-1β are encoded by a single gene and arise by alternative splicing. SDF-1α and SDF-1β are encoded by three and four exons, respectively[2].

Gene location

The human *SDF-1* gene is mapped to chromosome 10q by fluorescence *in situ* hybridization.

Amino acid sequence

Human (SDF-1α)

```
 1  MDAKVVAVLA LVLAALCISD GKPVSLSYRC PCRFFESHIA RANVKHLKIL
51  NTPNCALQIV ARLKNNNRQV CIDPKLKWIQ EYLEKALNK
```

Human (SDF-1β)

```
 1  MDAKVVAVLA LVLAALCISD GKPVSLSYRC PCRFFESHIA RANVKHLKIL
51  NTPNCALQIV ARLKNNNRQV CIDPKLKWIQ EYLEKALNKR LKM
```

Mouse

```
 1  MDAKVVAVLA LVLAALCISD GKPVSLSYRC PCRFFESHIA RANVKHLKIL
51  NTPNCALQIV ARLKNNNRQV CIDPKLKWIQ EYLEKALNK
```

Database accession numbers

	GenBank/EMBL	SwissProt	Ref
Mouse	D21072	P40224	3
Human	U16752	P48061	4

References
1. Nagasawa, T. et al. (1994) *Proc. Natl. Acad. Sci. USA* 91, 2305–2309.
2. Shirozu, M. et al. (1995) *Genomics* 28, 495–500.
3. Tashiro, K. et al. (1993) *Science* 261, 600–603.
4. Spotila, L.D. (1994) Direct GenBank submission.

MIG — Monokine induced by interferon gamma

Family

α family (CXC family)

Molecule

MIG was discovered by differential screening of a cDNA library prepared from lymphokine-activated macrophages. The *MIG* gene is inducible in macrophages and in other cells in response to IFN-γ.

Tissue sources[1]

Lymphokine activated macrophages, IFN-γ-treated human peripheral blood monocytes, IFN-γ-treated human monocytic cell line THP-1.

Target cells

Human tumor-infiltrating lymphocytes (TILs)[1], monocytes.

Physicochemical properties

Property	Human	Mouse
Signal	1–22	1–21
Amino acids		
Precursor	125	126
Mature	103	105
Disulfide bonds	a.a. 31 to 58	a.a. 30–57
	a.a. 33 to 74	a.a. 32–73
Glycosylation sites	1 (a.a. 58)	
Molecular weight		
Predicted	14019	14471

Transcription factors

The 5′ flanking region of the *MIG* gene contains an IFN-γ-responsive enhancer, γRE-1, consisting of an extended imperfect palindrome. A novel transcription factor γRF-1, which binds to the γRE-1 element, is rapidly activated in a variety of primary cell types and tumor cell lines treated with IFN-γ. One or more subunits of γRF-1 are antigenically related to p91/STAT1α. Activation of γRF-1 is unique, requiring both membrane and cytosol fractions and inhibition of endogenous tyrosine phosphatase activity[2,3].

Regulation of expression

IFN-γ induces the expression of MIG in a variety of monocytic cell lines as well as in macrophages.

In vitro biological effects

T lymphocytes

Transient elevation of $[Ca^{2+}]_i$ in human TILs and in cultured, IL-2-activated human peripheral blood-derived lymphocytes.
Chemotaxis.

Gene location

The human gene is located on chromosome 11.

Amino acid sequence

Human

```
1   MKKSGVLFLL GIILLVLIGV QGTPVVRKGR CSCISTNQGT IHLQSLKDLK
51  QFAPSPSCEK IEIIATLKNG VQTCLNPDSA DVKELIKKWE KQVSQKKKQK
101 NGKKHQKKKV LKVRKSQRSR QKKTT
```

Mouse

```
1   MKSAVLFLLG IIFLEQCGVR GTLVIRNARC SCISTSRGTI HYKSLKDLKQ
51  FAPSPNCNKT EIIATLKNGD QTCLDPDSAN VKKLMKEWEK KINQKKKQKR
101 GKKHQKNMKN RKPKTPQSRR RSRKTT
```

Database accession numbers

	GenBank/EMBL	SwissProt	PIR	Ref
Human	X72755	Q07325	JN0470	3
Mouse	M34815	P18340	A35766	2
	G199693			

References
1   Liao, F. et al. (1995) *J. Exp. Med.* 182, 1301–1314.
2   Farber, J.M. (1990) *Proc. Natl. Acad. Sci. USA* 87, 5238–5242.
3   Farber, J.M. (1993) *Biochem. Biophys. Res. Commun.* 192, 223–230.

MCP-1 — Monocyte chemoattractant protein-1

Alternate names

Monocyte chemotactic and activating factor (MCAF), lymphocyte derived chemotactic factor (LDCF), glioma derived chemotactic factor (GDCF), tumor derived chemotactic factor (TDCF), smooth muscle cell derived chemotactic factor (SMC-CF), JE gene product (mouse MCP-1)

Family

β family (CC family)

Molecule

MCP-1 is a prototypic β family chemokine with extensive characterization. MCP-1 was originally isolated as a product of the immediate early gene, *JE*, induced in response to PDGF. When the human homolog of *JE* was cloned, the encoded protein was shown to be identical to an authentic chemokine, MCP-1. MCP-1 is thought to be one of the most important chemokines for chronic inflammatory diseases that are controlled by mononuclear leukocytes.

Tissue sources

Fibroblasts[1], monocytes and macrophages, mouse spleen lymphocytes, endothelial cells, smooth muscle cells, cardiac myocytes, renal cortical epithelial cells, human epithelial cell line HEP-2, intestinal epithelial cells, Caco-2 cells, keratinocytes, mesangial cells, osteoblasts, liver fat storing cells, chondrocytes, melanocytes, luteal cells, mesothelial cells, bone marrow stromal cells, astrocytes, tumor cell lines such as malignant glioma, histiocytoma, astrocytoma, murine mast cell lines such as Cl.MC/9 and Cl.MC/C57.1, human osteosarcoma MG-63, lung carcinoma CALU-3, human sarcomas 8387 and KS.

Target cells

Monocytes[2] hematopoietic precursors, T lymphocytes, basophils, eosinophils, mast cells, NK cells, cardiac myocytes, osteoclasts, microglial cells, dendritic cells.

Physicochemical properties

Property	Human	Mouse	Guinea pig	Rat	Rabbit
pI (mature)	9.7	10.5	~10	~10	~10
Signal	1–23	1–23	1–23	1–23	1–23
Amino acids					
Precursor	99	148	120	148	125
Mature	76	125	97	125	102
Disulfide bonds	a.a. 34–59	34–59	33–57	34–59	34–59
	a.a. 35–75	35–75	34–73	35–75	35–75
Glycosylation sites	1	1 (a.a. 126)	0	1 (a.a. 126)	0

Property	Human	Mouse	Guinea pig	Rat	Rabbit
Molecular weight					
Precursor	11 025	16 326	13 741	16 460	13 776
Recombinant	8685	12 000			
Glycosylated	15 000	16 000–18 000	25 000		

Transcription factors

The MCP-1 gene promoter contains transcription factor binding sites for AP-1 (two sites, beginning at –128 and –156), octamer transcription factor (beginning at –282), and NFκB (beginning at –148). Inhibition of c-*fos* and c-*jun* inhibits MCP-1 gene expression.

Regulation of expression

Stimulus	Cell type	Response
Cytokines		
PDGF	Human or murine fibroblasts	Stimulation of MCP-1 mRNA expression
IFN-γ	Human keratinocytes	Up-regulation of MCP-1 constitutive expression
IL-1β and IFN-γ	Human fibroblasts	Production of MCP-1
	Human monocytes	Production of MCP-1
	Human umbilical vein cells	
	Human lung microvascular endothelial cells	
	Human renal cortical epithelial cells	
	Human epithelial cell line, HEP-2	Enhanced MCP-1 expression
	Human liver fat storing cells	Stimulation of constitutive expression
	Melanocytes	Increased MCP-1 expression
	Bone marrow derived murine stromal cell line +/+/–. LDA11	Induction of mRNA expression Expression can be potentiated by TGF-β1 or IL-4.
IL-1β and TNF-α	Human gingival fibroblasts	Stimulation of mRNA and protein expression
	Human pulmonary and fetal lung fibroblasts	
	Endometriotic fibroblast like cells	Induces time and dose dependent release of MCP-1
	Rat pulmonary alveolar macrophages	Expression of MCP-1 mRNA
	Human umbilical vein cells	Stimulation of mRNA expression
	Human lung microvascular endothelial cells	Stimulation of MCP-1 expression
	Human smooth muscle cells	Stimulation of MCP-1 production
	Cardiac myocytes	Stimulation of mRNA expression
	Neonatal rat cardiac cells	Stimulation of MCP-1 expression
	Human renal cortical epithelial cells	Expression of MCP-1 mRNA
	Human epithelial cell lines, A549 and At	Enhanced MCP-1 expression

Stimulus	Cell type	Response
IL-1β and TNF-α	Human epithelial cell line HEP-2	Enhanced MCP-1 expression
	Intestinal epithelial cells and Caco-2 cells	Enhanced constitutive expression
	Mesangial cells	Enhanced basal MCP-1 expression
	Chondrocytes	Enhanced MCP-1 mRNA expression
	Human pleural mesothelial cells	Induction of MCP-1 expression
	Murine astrocytes	Induction of MCP-1 production Pretreatment with IFN-γ augments production
	Human astrocytoma cell lines	Induction of MCP-1 mRNA and protein
	Human and rat glioma and fibrous histiocytoma	Enhanced constitutive expression of MCP-1
TNF-α, IL-1(α and β) and TGF-β	Human synovial fibroblasts	Expression of mRNA
	Human osteoblasts	Enhanced MCP-1 production
	Chondrocytes	Production of MCP-1
IL-1α	Human dermal fibroblasts	Rapid and transient increase in mRNA
GMCSF	Human monocytes or U937 cells	Enhancement of constitutive expression of MCP-1 mRNA
MCSF	Human or murine monocytes	Induction of MCP-1 mRNA and protein
	Endothelial cells	Stimulation of mRNA and protein production
IL-4	HUVECs	Up-regulation of MCP-1 mRNA

Bacteria

Streptococci	Human monocytes	Production of MCP-1
Mycobacterium tuberculosis[2]	Bone marrow derived murine macrophages	Expression of MCP-1 mRNA

Viruses

Measles virus	Human fibroblasts	Production of MCP-1
	Human osteosarcoma, MG-63	Enhanced expression of MCP-1

Others

PKC or PKA stimulation	Human synovial fibroblasts Monocytes	Enhanced MCP-1 production
	Chondrocytes	Stimulation of MCP-1 expression
Retinoic acid	Human synovial fibroblasts	Enhanced MCP-1 production
	IL-1 and LPS-stimulated chondrocytes	Potentiation of the effect of IL-1 and LPS
LPS	Human gingival fibroblasts	Stimulation of MCP-1 secretion
	Human PBMC	Induction of mRNA and protein expression
	Rat pulmonary alveolar macrophages	Induction of MCP-1 mRNA expression
	Rabbit spleen cells	Induction of MCP-1 mRNA
	Human umbilical vein cells	Stimulation of mRNA expression
	Human lung microvascular endothelial cells	Stimulation of MCP-1 expression
	Human renal cortical epithelial cells	Stimulation of MCP-1 expression
	Chondrocytes	Expression of MCP-1 mRNA and protein

Stimulus	Cell type	Response
PMA	Monocytes and THP-1 cells	Production of MCP-1
	HL-60 cells	Induction of MCP-1 mRNA expression
	Murine osteoblasts	Induction of MCP-1 expression Down-regulated by inhibition of PKC
PHA	Human PBMC	Induction of MCP-1 mRNA and protein expression
Purified protein derivative of *M. tuberculosis*	Human blood monocytes	Induction of MCP-1 mRNA
Serum	Human fibroblasts	Enhanced expression of mRNA and protein
	Mesangial cells	Induction of MCP-1 expression. Reduced by treatment of cells with lovastatin
In vitro culture	Human monocytes	Decrease in MCP-1 expression. It can be restored by IFN-γ or by cross-linking Fcγ receptors
Superoxide production	Human monocytes	Induction of MCP-1 production
Oxygen radicals	Mesangial cells	Induction of MCP-1 expression
Inhibitors of NADPH oxidase Free radical scavengers	TNF-stimulated human monocytes	Inhibition of MCP-1 mRNA expression
Poly(rI:rC)	Fibroblasts, monocytes	Induction of MCP-1 expression
Endothelin-1 production	Human monocytes	Induction of MCP-1 expression. Antioxidants inhibit this induction
Phagocytosis particulate stimuli (*M. tuberculosis*, latex beads or zymosan)	THP-1 or MonoMac6 cell lines	Induction of MCP-1 mRNA expression. The level of induction is independent of strain virulence or pre-exposure to TNF or HIV
Aggregated immunoglobulin, liposome encapsulated muramyl tripeptide and a synthetic peptide derived from C-reactive protein	Human monocytes	Induction of MCP-1 production
Adhesion and increased cell density	Human or murine monocytes	Expression of mRNA
ABO-incompatible red blood cells	Human peripheral blood leukocytes	Expression of MCP-1 mRNA
Con A	Rabbit spleen cells	Stimulation of MCP-1 mRNA expression
	Murine mast cell lines Cl.MC/9 and Cl.MC/C57.1	Stimulation of the basal expression
Minimally oxidized LDL	Human umbilical vein cells	Induction of MCP-1 mRNA and protein
	Smooth muscle cells	Induction of mRNA and protein
	Mesangial cells	Induction of MCP-1 mRNA expression

Stimulus	Cell type	Response
Thrombin	Human and rodent smooth muscle cells	Stimulation of MCP-1 mRNA expression. The protease inhibitor hirudin blocks this stimulation
	Vascular pericytes	Stimulation of MCP-1 expression. This effect requires both catalytic activity and receptor binding.
	Human liver fat storing cells	Increased expression of MCP-1 mRNA and protein
Hypoxia	Cardiac myocytes	Induction of MCP-1 mRNA expression
IgG immune complexes	Murine mesangial cells	Increased transcription of MCP-1 mRNA. This action is mediated by activation of NADPH oxidase
Drugs		
Dexamethasone	TNF-α and IL-1α stimulated human fibroblasts	Inhibition of MCP-1 production
	Monocytes	Inhibition of MCP-1 production
	Intestinal epithelial cells and Caco-2 cells	Suppression of IL-1 induced MCP-1 expression
	Chondrocytes	Inhibition of MCP-1 production
Steroids	Human and rat glioma and fibrous histiocytoma	Inhibition of constitutive expression
	Human mast cell line, HMC-1	Down-regulation of PMA-induced MCP-1 mRNA

[a] Constitutive expression of MCP-1 mRNA is seen in normal cultures of human luteal cells and tumor cell lines such as human sarcomas 8387 and KS, human lung carcinoma CALU-3; a highly metastatic human melanoma cell line TXM-1 expresses MCP-1 mRNA.
[b] Human monocytes do not produce MCP-1 in response to IL-4, IL-1β or TNF-α under serum-free conditions.

Expression in disease

Disease	Tissue	Comment
Human disease		
Chronic active hepatitis, fulminant hepatic failure	Liver tissue	Abundant expression of MCP-1 mRNA
Primary and metastatic human melanoma	Skin	Increased MCP-1 expression as shown by immunohistochemistry and *in situ* hybridization
RA	Synovial macrophages, vascular endothelial cells and synoviocytes	Constitutive expression as shown by immunohistochemistry
RA	Non-T-cell populations of synovial fluid and synovial tissue	Increased MCP-1 expression, but absent in peripheral blood
RA and osteoarthritis	Synovial lining layer	High levels of MCP-1 protein expression
Human atherosclerotic lesions[3]	Endothelial cells and macrophage rich regions	MCP-1 mRNA and protein present
Idiopathic dilated cardiomyopathy	Cardiac biopsy	MCP-1 mRNA expression
Ovarian carcinomas	Epithelial areas	MCP-1 mRNA and protein expression

Disease	Tissue	Comment
Pregnancy	Human endometrial tissue and decidua parietalis	MCP-1 mRNA expression in mid trimester and at term. No relationship to day of cycle, endocrine status or duration of pregnancy
Meningiomas	Tumor tissue	Increased expression of mRNA and protein associated with macrophage infiltration and perifocal edema
Human malignant glioma	Glioma cells with large and pleomorphic nuclei, tumor vessels	MCP-1 mRNA and protein expression. Degree of macrophage infiltration correlated with the level of mRNA expression
Brain tumors	Glioblastoma, astrocytoma grade 2, ependymoma and medulloblastoma, neoplastic astrocytes and tumor cyst fluids	MCP-1 mRNA and protein expression
Lichenoid dermatitis, dermal hypersensitivity and spongiotic dermatitis	Endothelial cells of dermal microvessels and infiltrating macrophages	MCP-1 expression
Psoriasis	Proliferating basal keratinocytes of the tips of the rete ridges	MCP-1 mRNA expression
Glomerulonephritis, idiopathic crescentic proliferative glomerulonephritis, Wegener's granulomatosis and systemic lupus erythematosus	Glomeruli	MCP-1 detected
Ulcerative colitis or Crohn's disease	Mucosal tissue	In normal gut mucosa, MCP-1 mRNA is detected in surface epithelium. Inflamed mucosa contains multiple positive cells including spindle cells, mononuclear cells and endothelial cells
Periodontal disease	Gingival tissue and crevicular fluid Vascular endothelial cells, fibroblasts and macrophages	MCP-1 mRNA and protein expression in parallel with phagocyte infiltration
Idiopathic pulmonary fibrosis	Pulmonary epithelial cells, monocytes, vascular endothelial and smooth muscle cells	Strong expression of MCP-1 mRNA and protein
	Pulmonary fibroblasts	Spontaneous production of MCP-1
Allergic rhinitis	Lamina propria of allergic nasal mucosa around small vessels and excretory glands	MCP-1 present

Disease	Tissue	Comment
Summer type hypersensitivity pneumonitis	Bronchoalveolar lavage fluid	MCP-1 levels elevated
Asthma	Bronchial biopsies containing bronchial epithelium, subepithelial macrophages, blood vessels and bronchial smooth muscle cells	Strong expression of MCP-1
Sarcoidosis	Bronchoalveolar lavage fluid	Significant increase in MCP-1 protein

Animal models of disease

Disease	Tissue	Comment
Bleomycin-induced pulmonary fibrosis	Eosinophils, alveolar macrophages	Elevation of MCP-1 mRNA and protein
Rat model of glucan-induced pulmonary granulocytosis	Lungs	Expression of MCP-1
Mice exposed to aerosols of *M. tuberculosis*	Lungs	Expression of MCP-1 mRNA
Immune complex induced lung injury in rats	Pulmonary alveolar macrophages	Expression of MCP-1 mRNA
Rabbit atherosclerosis	Atherosclerotic lesions and foam cells isolated from arterial lesions	MCP-1 mRNA present
Primate model of atherosclerosis	Smooth muscle cells of the medial layer of the artery, macrophages and smooth muscle cells in the overlying intimal lesion	Strong expression of MCP-1 mRNA
Rat model of ischemia	Heart tissue	mRNA detected
Venous thrombosis in rats	Vein wall homogenates	MCP-1 protein detected
Rat model of cerebral ischemia	Ischemic brain tissue containing macrophages and endothelial cells	MCP-1 detected beginning 6 h following middle cerebral artery occlusion; suppressed by steroid treatment
Murine model of experimental autoimmune encephalomyelitis	Spinal cord and brain parenchymal cells resembling astrocytes	Expression of MCP-1 during the course of acute stage in correlation with the onset of histologic and clinical disease
Rats with unilateral urethral obstruction	Apical segments of cortical tubules of kidneys	MCP-1 mRNA and protein detected
Experimental glomerulonephritis in rats	Kidneys	Transient MCP-1 up-regulation in the early stages
Mouse model of anti-GBM glomerulonephritis	Nephrotic glomeruli	Expression of MCP-1 mRNA
Rat model of protein overload proteinuria	Kidneys	Elevation of MCP-1 mRNA levels
Rabbit knee joints exposed to polyethylene particles	Mononuclear cells in the vicinity of the particle	Elevation of MCP-1 mRNA levels

Disease	Tissue	Comment
Rats exposed to hepatotoxins	Fat storing or Ito cells of liver	Expression of MCP-1 mRNA and protein
Rat model of chronic rejection of renal allografts	Kidneys	Dramatic increase in MCP-1 mRNA preceding a strong macrophage infiltration
Rat model of cardiac allografts	Macrophages in rejecting allografts	Expression of MCP-1 mRNA and protein
Mice infected with BCG or *listeria*	Murine liver Kupffer cells and infiltrating macrophages	Expression of MCP-1 mRNA
E. coli infected baboons	Plasma	Increased MCP-1 within 2 h after intravenous injection
Rat model of uveitis	Uvea and retina but not cornea	MCP-1 mRNA is seen before the onset of clinical uveitis

[a] Normal human serum and urine contains detectable levels of MCP-1 protein with a mean concentration of 138 pg/ml for serum and 158 pg/ml for urine. There is no difference in the levels between males and females and no correlation with age.

In vivo studies

Animal studies

Murine colon carcinoma cells transfected with murine MCP-1 produce fewer lung metastases.

The growth and metastasis of the renal adenocarcinoma cell line RENCA is reduced when admixed with syngeneic fibroblasts expressing protein.

Transfected murine myeloma cells producing human MCP-1 protein show greater tumor necrosis and infiltration of macrophages into the tissue surrounding implanted tumors. The macrophages express MCP-1 mRNA.

Human MCP-1 injected into rabbit joints induces a marked infiltration of macrophages.

Injection of human MCP-1 protein into canine dermis causes a mild perivascular cuffing of monocytes.

Injection of human MCP-1 protein into the hind flanks of SCID mice which have been engrafted with human lymphocytes induces significant migration of human lymphocytes to the injection site by 4 h.

Injection of human protein intradermally in rabbits induces an infiltration of monocytes at 18 h. The infiltration of cells is characterized by monocyte clustering and adherence to endothelium.

Administration of human MCP-1 protein by intracerebroventricular infusion into rats decreases short-term food intake.

Administration of guinea pig MCP-1 protein intradermally into guinea pig skin induces a strong macrophage infiltration at the injection site after 6 h.

Administration of human MCP-1 protein to mice prior to infection with *Pseudomonas* or *Salmonella* enhances resistance to infection. Peritoneal exudate cells from treated mice exhibit enhanced phagocytosis and killing of bacteria.

Sections of xenografted tumors from the CALU-3 human lung carcinoma in mice demonstrate MCP-1 protein by immunohistochemistry. Human peripheral blood leukocytes injected in the vicinity of the graft infiltrate and induce tumor necrosis.

In vitro biological effects

Monocytes

Mobilization of intracellular free calcium (0.3–300 nM)
Mobilization of intracellular free calcium – cell lines (10–100 nM)
Generation of respiratory burst (0.1 nM–1 μM)
Expression of β_2 adhesion molecules (1 nM–1 μM)
Chemotaxis (0.01–100 nM)[4,5]
Chemotaxis – THP-1 cells (0.1–100 nM)
Induction of cytokines (0.1–20 nM)
Degranulation (0.5 nM)
Enhanced tumoricidal activity (0.01–10 nM)
Enzyme release (1–100 nM)
Inhibition of adenylate cyclase (0.03–100 nM)
Release of arachidonic acid (1–10 nM)

Hematopoietic precursors

Inhibition of proliferation (0.25–10 nM)

T lymphocytes

Mobilization of intracellular free calcium (0.05–50 nM)
Chemotaxis (0.01 nM–1 μM)

Basophils

Chemotaxis (1–100 nM)
Release of histamine (0.5 nM–1 μM)
Mobilization of intracellular free calcium (5 nM)

Eosinophils

Mobilization of intracellular free calcium (100 nM)

Mast cells

Chemotaxis (1–100 nM)

Natural killer cells

Chemotaxis (0.01 – 0.1 nM)
Chemotaxis, activated cells (0.1–20 nM)

Cardiac myocytes

Induction of ICAM-1 (10–100 nM)

Osteoclasts

Chemotaxis (0.5–10 nM)

Dendritic cells

Chemotaxis (10 nM)

Microglial cells

Chemotaxis (0.1–10 nM)

Intracellular signaling

Causes the activation of GTPase by functional interaction with the G-protein coupled receptors on monocytes. This activation is sensitive to ADP-ribosylating PT[6].

Causes inhibition of adenylate cyclase and decreases accumulation of cAMP in intact responsive cells.

Causes PT-sensitive influx of extracellular Ca^{2+} which results in a rapid and transient increase of intracellular Ca^{2+} concentration.

Causes breakdown of phosphatidylinositol bisphosphate and accumulation of inositol phosphates.

MCP-1-induced Ca^{2+} influx is necessary but not sufficient for arachidonic acid accumulation.

Activation of phospholipid A_2 is an important step in the chemotactic signaling by MCP-1.

Inhibitors of GMP-dependent kinases and myosin L-chain kinases have no effect on MCP-1-induced Ca^{2+} influx in monocytes, while inhibitors of PKC/cAMP dependent kinase such as H-7, HA-1004, and staurosporin markedly decrease MCP-1-induced Ca^{2+} influx in monocytes.

Receptor binding characteristics

Binds to CCR1, CCR2, DARC and US28

Cross-desensitization

MCP-1 desensitizes monocytes for subsequent activation by MCP-3, MIP-1α, and RANTES in calcium mobilization and chemotaxis. MCP-1 desensitizes enzyme release by MCP-2 and MCP-3 and to a lesser extent by RANTES and MIP-1α. The MCP-1 response is desensitized by prior exposure to MCP-2 and MCP-3 but not RANTES or MIP-1α.

The calcium response in human monocytes is not desensitized by prior exposure to RANTES or MIP-1α but prior exposure to MCP-1 desensitizes a response to MIP-1α or RANTES.

MCP-1 desensitizes basophils and eosinophils for subsequent calcium mobilization by MCP-3 but not RANTES and MIP-1α. MIP-1α, MCP-3, and

RANTES do not desensitize monocytes, basophils, or eosinophils for subsequent stimulation by MCP-1.

In basophil mediator release, MCP-2 and MCP-3 can desensitize a subsequent response to MCP-1.

MCP-1 desensitizes T lymphocytes for calcium mobilization by MCP-2 and MCP-3.

MIP-1α desensitizes THP-1 cells for calcium response to MCP-1 but MCP-1 does not desensitize a subsequent MIP-1α response.

Gene structure

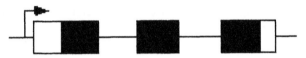

MCP-1 is a single copy gene in human, monkey, bovine, pig, rat, and mouse.

Human

Gene is configured in three exons of 145, 118, and 478 bp with two introns of 800 and 385 bp. There is a CAAT box at −97 and no TATA box.

Murine

The gene has three exons and when spliced the mRNA is either 594 or 797 bases. There is a candidate TATA box at −30.

Gene location

The human *MCP-1* gene, designated *Scya2*, is located on human chromosome 17 (q11.2–q21.1) in humans. The murine *JE* gene is located on the distal portion of murine chromosome 11 near the *Hox-2* gene complex.

Protein structure

By analogy to the structure of IL-8, the MCP-1 monomer consists of a triple stranded antiparallel β-sheet (residues 26–31, 39–44, and 48–52) arranged in a Greek key, on the top of which lies an α-helix (residues 57–70). Forms dimers with a dissociation constant of 100 nM. Amino acids 3–9 form the putative dimer interface. Synthetic peptide 13–35 stimulates monocyte chemotaxis and competes for MCP-1 binding to receptors. Residues 7–10 are essential for desensitization but are not sufficient for function; residues 1–6 are required for function. Truncation analogs 8–76, 9–76, and 10–76 are MCP-1 antagonists. Exists as a monomer in solution at maximally active *in vitro* concentrations[8].

Amino acid sequence

Human[9]

```
MKVSAALLCL LLIAATFIPQ GLAQPDAINA PVTCCYNFTN RKISVQRLAS
YRRITSSKCP KEAVIFKTIV AKEICADPKQ KWVQDSMDHL DKQTQTPKT
```

Murine (JE)[10]

```
MQVPVMLLGL LFTVAGWSIH VLAQPDAVNA PLTCCYSFTS KMIPMSRLES
YKRITSSRCP KEAVVFVTLL KREVCADPKK EWVQTYIKNL DRNQMRSEPT
TLFKTASALR SSAPLNVKLT RKSEANASTT FSTTTSSTSV GVTSVTVN
```

Rat

```
MQVSVTLLGL LFTVAACSIH VLSQPDAVNA PLTCCYSFTG KMIPMSRLEN
YKRITSSRCP KKLVVFVTKL KREICADPNK EWVQKYIRLK DQNQVRSETT
VFYKIASTLR TSAPLNVNLT HKSEANASTL FSTTTSSTSV EVTSMTEN
```

Rabbit

```
MKVSATLLCL LLIAVAFSSH VLAQPDAVNS PVTCCYTFTN KTISVKRLMS
YRRINSTKCP KEAVIFMTKL AKGICADPKQ KWVQDAIANL DKKMQTPKTL
TSYSTTQEHT TNLSSTRTPS TTTSL
```

Bovine (MCP-1α)

```
MKVSAALLCL LLTVAAFSTE VLAQPDAINS QVACCYTFNS KKISMQRLMN
YRRVTSSKCP KEAVIFKTIL GKELCADPKQ KWVQDSINYL NKKNQTPKP
```

Bovine (MCP-1β)

```
QPDAINSPVT CCYTLTSKKI SMQRLMSYRR VTSSKCPKEA VIFKTIAGKE
ICAEPKWVQD SISHLDKKNQ XPKP
```

Guinea pig

```
MQRSSVLLCL LVIEATFCSL LMAQPDGVNT PTCCYTFNLQ IPLKRVKGYE
RITSSRCPQE AVIFRTLKNK EVCADPTQKW VQDYIAKLDQ RTQQKQNSTA
PQTSKPLNIR FTTQDPKNRS
```

Porcine

```
MKVSAALLCL LLTAATFCTQ VLAQPDAINS PVTCCYTLTS KKISMQRLMS
YRRVTSSKCP KEAVIFKTIA GKEICAEPKQ KWVQDSISHL DKKNQTPKP
```

Database accession numbers

Species	SwissProt	EMBL/GenBank	PIR	Ref
Human	P13500	M31626	A35474	16
		M30816	S03339	
		M31625		
		M24545		
		M28226		
		X14768		
		M37719		
		M28223		
		M28224		
Murine	P10148	J04467	A30209	9
		M19681	A30861	
			S16226	
Rat	P14844	X17053	JN0128	11
		M57441	S07723	
Rabbit	P28292	M57440		13
Bovine α	P28291	L32659	A39296	14
		M84602	JC2336	
Bovine β	P80343			
Guinea pig	Q08782	L04985		12
Porcine	P42831	Z48479		15
		S69211		
		X79416		

References

1 Rollins, B.J. (1991) *Cancer Cells* 3, 517–524.
2 Leonard, E.J. and Yoshimura, T. (1990) *Immunology Today* 11, 97–101.
3 Valente, A.J. et al. (1990) *Circulation* 86, III-20–III-25.
4 Zachariae, C.O.C. et al. (1990) *J. Exp. Med.* 171, 2177–2182.
5 Mantovani, A. et al. (1995) *Reg. Immunol.* 6, 445–448.
6 Locati, M. et al. (1995) *J. Immunol.* 269, 4746–4753.
7 Van Riper, G. et al. (1993) *J. Exp. Med.* 177, 851–856.
8 Paolini, J.F. et al. (1994) *J. Immunol.* 153, 2704–2717.
9 Rollins, B.J. et al. (1988) *Proc. Nat. Acad. Sci. USA* 85, 3738–3742.
10 Gong, J.-H. and Clark-Lewis, I. (1995) *J. Exp. Med.* 181, 631–640.
11 Yoshimura, Y., Takeya, M. and Takahashi, K. (1991) *Biochem. Biophys. Res. Commun.* 174, 504–509.
12 Yoshimura, T. (1993) *J. Immunol.* 150, 5025–5032.
13 Yoshimura, T. and Yuhki, N. (1991) *J. Immunol.* 146, 3483–3488.
14 Wempe, F. et al. (1991) *DNA Cell Biol.* 10, 671–679.
15 Hosang, K. et al. (1994) *Biochem. Biophys. Res. Commun.* 199, 962–968.
16 Yoshimura, T. et al. (1989) *FEBS Lett.* 244, 487–493.

MCP-2 — Monocyte chemoattractant protein-2

Alternate names

c11/1 (bovine)

Family

β family (CC family)

Molecule

Originally reported as a variant of MCP-1, MCP-2 has been identified as an independent chemokine with several distinct biological properties such as activation of eosinophils and basophils. The main differences between MCP-1 and MCP-2 is in the N-terminal region of the molecule.

Tissue sources

Fibroblasts, peripheral blood mononuclear cells, luteal cells[1], osteosarcoma cell line MG 63

Target cells

T lymphocytes, monocytes, eosinophils, basophils, NK cells

Physicochemical properties

Property	Human	Bovine	Porcine
pI (mature)	9.7	~9	~9
Signal	?	1–23	1–23
Amino acids			
Precursor	77	99	99
Mature	77	76	99
Disulfide bonds	a.a. 11–36	34–58	
	a.a. 12–50	35–75	
Glycosylation sites	0	0	
Molecular weight			
Precursor	7500	10 900	

Transcription factors

The MCP-2 promoter contains a TATAA site and a putative AP-1 consensus site[2].

Regulation of expression

Human diploid fibroblasts produce MCP-2 protein in response to addition of IL-1β, IFN-γ, polyrI:rC, or measles virus.

MCP-2 protein is produced by human mononuclear cells in response to addition of IL-1β, IFN-γ, polyrI:rC, Con A, or measles virus.

MCP-2 mRNA is found in bovine peripheral blood leukocytes following stimulation with PHA.

MCP-2 mRNA is expressed in normal leuteal cell cultures.

MCP-2 protein and mRNA are expressed in the osteosarcoma cell line MG-63 after addition of IL-1β, IFN-γ, or measles virus.

Expression in disease

Mononuclear cells from bronchoalveolar lavage, but not peripheral blood, from an allergic patient express MCP-2 mRNA upon *in vitro* exposure to allergen[1].

In vivo studies

Injection of human MCP-2 protein intradermally in rabbits induces an infiltration of monocytes at 18 h. The infiltration of cells is characterized by monocyte clustering and adherence to endothelium[2].

In vitro biological effects
T Lymphocytes

Chemotaxis (0.01 nM–1 μM)
Mobilization of intracellular free calcium (5–50 nM)

Monocytes[3]

Chemotaxis (0.01–100 nM)
Enzyme release (3–100 nM)

Eosinophils[4]

Chemotaxis (10–100 nM)

Basophils[5]

Histamine release (100 nM)
Chemotaxis (10–100 nM)
Mobilization of intracellular free calcium (100 nM)

Natural killer cells (activated)

Chemotaxis (3–20 nM)

Cross-desensitization

Response to MCP-2 in lymphocytes can be desensitized by prior exposure to MCP-1 or MCP-3.

In eosinophils, MCP-2 chemotaxis can be desensitized by prior exposure to RANTES or MCP-3 but not MCP-1. MCP-2 can desensitize a chemotactic response of eosinophils to RANTES. MCP-2 can desensitize a calcium response to MCP-3 but not MIP-1α or RANTES while the calcium response to MCP-2 can be desensitized by MCP-3 and RANTES but not MIP-1α.

In human basophils, MCP-2 can desensitize a calcium response to MCP-3, MIP-1α or RANTES while the calcium response to MCP-2 can be desensitized by MCP-3, MCP-1 and, to a lesser extent, by RANTES. MCP-2 can desensitize mediator release induced by MCP-1 or MCP-3.

In human monocytes, MCP-2 desensitizes enzyme release by MCP-1, MCP-3, and RANTES and to a lesser extent by MIP-1α. The MCP-2 response is desensitized by prior exposure to MCP-1 and MCP-3, to a lesser extent by RANTES and not at all by MIP-1α.

Gene structure

Human

A single copy gene is present in porcine and bovine systems. The bovine gene consists of three exons: exon 1 of 149 bp, exon 2 of 118 bp and exon 3 of 555 bp. Exon 3 contains the putative polyadenylation signal[6].

Gene location

The human *MCP-2* gene is located on chromosome 17.

Amino acid sequence

Human[7]

```
QPDSVSIPIT CCFNVINRKI PIQRLESYTR ITNIQCPKEA VIFKTGKEVC
ADPKERWVRD SMKHLDQIFQ NLKP
```

Porcine[8]

```
MQVSAALLCL LLTTAAFSTQ VLAQPDSVSI PITCCFGLVN GKIPFKKLES
YTRITNSQCP QEAVIFKTKA DKEVCADPQQ KWVQNSMKLL DQKSQTPKP
```

Bovine

```
MKVSAGILCL LLVAATFGTQ VLAQPDSVST PITCCFSVIN GKIPFKKLDS
YTRITNSQCP QEAVIFKTKA DRDVCADPKQ KWVQTSIRLL DQKSRTPKP
```

Database accession numbers

Species	SwissProt	EMBL/GenBank	Ref
Human	P80075		9
Porcine		Z48480	8
Bovine	Q09141	S67954	2
		S67956	

References
[1] Alam, R. et al. (1994) *J. Immunol.* 153, 3155–3159.
[2] Wempe, F. et al. (1994) *DNA and Cell Biol.* 13, 1–8.
[3] Van Damme, J. et al. (1992) *J. Exp. Med.* 176, 59–65.
[4] Wuyts, A. et al. (1994) *J. Immunol. Methods* 174, 237–247.
[5] Noso, N. et al. (1994) *Biochem. Biophys. Res. Comm.* 200, 1470–1476.
[6] Weber, M. et al. (1995) *J. Immunol.* 154, 4166–4172.
[7] Proost, P. et al. (1995) *Cytokine* 7, 97–104.
[8] Hosang, K. et al. (1995) *Biochem. Biophys. Res. Comm.* 205, 148–153.
[9] Chang, H.C. et al. (1989) *Int. Immunol.* 1, 388–397.

MCP-3 | Monocyte chemoattractant protein-3

Alternate names

FIC (murine), MARC (murine), NC28

Family

β family (CC family)

Molecule

MCP-3 is rapidly becoming a distinct member of the β-chemokine family with very different binding and functional characteristics from MCP-1. MCP-3 binds to receptors on monocytes that are unique as well as shared by MCP-1. MCP-3, unlike MCP-1, is chemotactic to eosinophils and more potent toward basophils. MCP-3 is the only β-chemokine that does not form dimers at high concentrations.

Tissue sources

Fibroblasts[1], platelets, osteosarcoma MG-63[2], U937 cells, mast cells, and monocytes.

Target cells

Monocytes, T lymphocytes, basophils, eosinophils, NK cells (activated), dendritic cells, neutrophils.

Physicochemical properties

Property	Human	Murine
pI (mature)	9.7	~10
Signal	1–23	1–23
Amino acids		
Precursor	99	97
Mature	76	74
Disulfide bonds	a.a. 34–59	34–58
	a.a. 35–75	35–75
Glycosylation sites	1 (a.a. 29)	1 (a.a. 29)
Molecular weight		
Predicted	11 200	10 999
SDS-PAGE	13 000	11 000

Regulation of expression

Serum stimulated NIH 3T3 cells express MCP-3 mRNA[1].

Human platelets express MCP-3 mRNA constitutively.

MCP-3 protein and mRNA are expressed in the osteosarcoma cell line MG-63 after addition of IL-1β[2].

The U937 cell line expresses MCP-3 mRNA after treatment with phorbol esters.

The murine mast cell line CPII, stimulated with phorbol esters in combination with ionomycin or with dinitrophenol in combination with anti-dinitrophenol IgE antibodies expresses MCP-3 mRNA.

Analysis of unstimulated CPII cells or purified mouse peritoneal mast cells reveals basal MCP-3 transcription which is increased 2 h after IgE plus antigen challenge.

The major source of MCP-3 mRNA from peripheral blood cells is the adherent mononuclear cells corresponding to monocytes. MCP-3 expression is enhanced by LPS, IFN-γ, or PHA, and inhibited by IL-13.

Expression in disease

MCP-3 mRNA is expressed in allergen challenged skin sites in atopic subjects.

In vivo studies

Animal studies

Injection of human MCP-3 protein intradermally in rabbits induces an infiltration of monocytes at 18 h. The infiltration of cells is characterized by monocyte clustering and adherence to endothelium.

In vitro biological effects

Monocytes

Chemotaxis (0.01–100 nM)
Mobilization of intracellular free calcium (1–100 nM)
Enzyme release (1–100 nM)
Release of arachidonic acid (5 nM)

T lymphocytes

Chemotaxis (0.01 nM–1 µM)
Mobilization of intracellular free calcium (0.5–50 nM)

Eosinophils[3]

Chemotaxis (0.1–100 nM)

Basophils

Histamine release (10–100 nM)
Chemotaxis (1–100 nM)

NK cells (activated)

Chemotaxis (3–20 nM)

Dendritic cells

Chemotaxis (10 nM)
Mobilization of intracellular free calcium (10 nM)

Neutrophils[4]

Chemotaxis (0.01 nM–1 μM)

Intracellular signaling

Causes transient changes in intracellular calcium in monocytes, eosinophils, and basophils.

Causes potent inhibition of adenlyate cyclase in cells transfected with the MCP-1 receptor.

Receptor binding studies

Binds to CCR2, CCR3 and an unidentified receptor on monocytes, eosinophils and IL-2 activated NK cells.

Cross-desensitization

Calcium mobilization to MCP-3 in human lymphocytes can be desensitized by prior exposure to MCP-1 and prior MCP-3 exposure can desensitize a calcium response to MCP-2 but not MCP-1.

In human eosinophils, chemotaxis to RANTES can be desensitized by prior exposure to MCP-3 and vice versa. MCP-1 has no effect on this MCP-3 response. MCP-3 can cross-desensitize a calcium response to MCP-2.

In human monocytes, MCP-3 desensitizes enzyme release by MCP-1, MCP-2, MIP-1α, and RANTES. The MCP-3 response is desensitized by prior exposure to MCP-1 and MCP-2, to a lesser extent by RANTES, and not at all by MIP-1α.

In human monocytes, calcium mobilization to MCP-3 can be cross-desensitized by MCP-1 but not MCP-2.

In basophils, MCP-3 can cross-desensitize a calcium response to MCP-2.

Gene structure

Human

The MCP-3 gene is constructed as three exons with intervening introns. Two tandem dinucleotide repeat DNAs or microsatellites are present in the promoter region. The gene features CAAT, TATA, and CAP sites while a UA rich destabilizing sequence and an mRNA hairpin loop are located at the 3'-end.

Gene location

A single copy gene, termed Scya7, is found on chromosome 17 (q11.2–q12) in humans[5].

Amino acid sequence

Human[6]

```
MKASAALLCL LLTAAAFSPQ GLAQPVGINT STTCCYRFIN KKIPKQRLES
YRRTTSSHCP REAVIFKTKL DKEICADPTQ KWVQDFMKHL DKKTQTPKL
```

Murine[7]

```
MRISATLLCL LLIAAAFSIQ VWAQPDGPNA STCCYVKKQK IPKRNLKSYR
RITSSRCPWE AVIFKTKKGM EVCAEAHQKW VEEAIAYLDM KTPTPKP
```

Database accession numbers

Species	SwissProt	EMBL/GenBank	PIR	Ref
Human	P80098	X72308	JC1478	6
		X72309	S32222	
		X71087	A54678	
Murine	Q03366	Z12297	S30592	7
		L04694		
		S71251		

References

1 Heinrich, J.N. et al. (1993) *Mol. Cel. Biol.* 13, 2020–2030.
2 Opdenakker, G. et al. (1993) *Biochem. Biophys. Res. Commun.* 191, 535–542.
3 Dahinden, C.A. et al. (1994) *J. Exp. Med.* 179, 751–756.
4 Xu, L.L. et al. (1996) *Eur. J. Immunol.* 25, 2612–2617.
5 Opdenakker, G. et al. (1994) *Genomics* 21, 403–408.
6 Minty, A. et al. (1993) *European Cytokine Network* 4, 99–110.
7 Thirion, S. et al. (1994) *Biochem. Biophys. Res. Commun.* 201, 493–499.

MIP-1α — Macrophage inflammatory protein-1α

Alternate names

GOS19, LD78, pAT464 gene product, TY5 (murine), SISα (murine)

Family

β family (CC family)

Molecule

MIP-1α has a wide range of biological activities that include prostaglandin-independent pyrogenic activity, a potential role in wound healing, monocyte chemotaxis and suppression of immature bone marrow stem and progenitor cells. The most important finding about MIP-1α is perhaps the recent report of its HIV-suppressive effect.[13]

Tissue sources

Fibroblasts, monocytes[1], lymphocytes, neutrophils, eosinophils, smooth muscle cells, mast cells, platelets, bone marrow stromal cells, glial cells, epithelial cell line A549, progenitors of dendritic Langerhans' cells U937, Jurkat, K562[5].

Target cells

T lymphocytes, basophils, hematopoietic precursor cells, monocytes, eosinophils, neutrophils, mast cells, NK cells, epidermal keratinocytes, astrocytes, spermatogonia, dendritic cells, osteoclasts.

Physicochemical properties

Property	Human	Murine
pI (mature)	4.6	4.6
Signal	1–22	1–23
Amino acids		
Precursor	92	92
Mature	70	69
Disulfide bonds	a.a. 33–57	34–57
	a.a. 34–73	35–73
Glycosylation sites	1 (a.a. 29)	1 (a.a. 29)
Molecular weight		
Predicted	10085	10345
SDS-PAGE	16000	8000

Transcription factors

The promoter of the human MIP-1α gene contains five major nuclear protein binding sites for C/EBP, NF-κB, and/or c-ets family members. Changes in promoter binding by members of the C/EBP and NF-κB families correlate with transcriptional up-regulation.

The transcription factor ICK-1 acts as a strong negative regulator in Jurkat and K562, but not U937, cells. An additional binding site (–113 to –97), which overlaps the ICK-1 site (–106 to –94), binds a novel nuclear protein expressed in hematopoietic cells.

A comparison of the promoters for murine MIP-1β and murine MIP-1α reveals a conserved CK-1 element, but transient expression studies in RAW 264.7 macrophages with proximal fragments of either the MIP-1α or the MIP-1β 5' promoter fused to a human growth hormone reporter gene link LPS-inducibility in both to promoter segments near to, but not identical with, the consensus CK-1 sequence. Proximal 5' promoter fragments cloned from both the MIP-1α and MIP-1β genes confer constitutive expression on the fused reporter gene sequences in macrophage-like cells, but initial 5' deletion analysis does not link this responsiveness to known sequence motifs. The MIP-1α promoter is constituvely active in the myelomonocytic cell line WEHI 3B(A)d.

Regulation of expression

Stimulus	Cell type	Response
Cytokines		
IL-1	Human pulmonary fibroblasts	Induction of MIP-1α expression
IL-1, TNF-α, LPS	Synovial fibroblasts	Induction of MIP-1α production
	Human blood cells	Stimulation of MIP-1α production. With all agents, production can be suppressed by addition of IL-4 through destabilization of MIP-1α mRNA
	Rat alveolar macrophage cell line, NR8383	Induction of MIP-1α expression
	Human granulocytes	Induction of MIP-1α expression. This induction can be suppressed by addition of IL-10
	Epithelial cell line A549	Enhanced constitutive expression of MIP-1α
TGF-β	Bone marrow derived murine macrophages	Inhibition of MIP-1α mRNA expression
IFN-γ	Human neutrophils	Induction of MIP-1α mRNA expression
	Bone marrow stromal cells	Enhanced production of MIP-1α
GM-CSF	Human neutrophils	Augments LPS induced expression of MIP-1α by stabilizing mRNA
IL-1, TNF-α, IL-4, IL-10	Human pulmonary smooth muscle cells	Induction of MIP-1α mRNA expression
Bacteria		
Mycobacterium tuberculosis	Bone marrow derived murine macrophages	Induction of MIP-1α mRNA expression
Viruses		
HIV	Alveolar macrophages	Spontaneous production of MIP-1α
HIV-*tat* protein	Raji B cells	Induction of MIP-1α mRNA expression
	Jurkat T cell line	Down-regulation of MIP-1α stimulation by phorbol esters with PHA

Stimulus	Cell type	Response
HTLV-1	MT4 cells	Expression of MIP-1α protein
Others		
PMA[1]	Human fibroblasts	Enhancement of MIP-1α production
	HL-60 and U937 cells	Enhancement of MIP-1α mRNA expression[a]
	Peripheral blood lymphocytes (CD8+CD45RO+ population)	Induction of MIP-1α production
	Human tonsillar lymphocytes	Induction of MIP-1α production
	Human mast cell line HMC-1	Induction of MIP-1α production
	Human glioma cell line U105MG, but not U251MG	Induction of MIP-1α production
	BV-2 murine microglial cell line and murine astrocytes	Induction of MIP-1α mRNA. Expression is inhibited by addition of dibutyryl cAMP IFN-γ or PGE$_1$. Dexamethasone decreases mRNA in astrocytes but not BV-2 cells
PHA	Peripheral blood monocytes	Induction of MIP-1α production
	Lymphocytes	Induction of MIP-1α production
	Human tonsillar lymphocytes	Induction of MIP-1α production
	Raji B cell line	Induction of MIP-1α production in association with PMA
Con A	Murine Th1 and Th2 cell lines	Induction of MIP-1α production
	Murine mast cell lines Cl.MC/9, Cl.MC/9.2 and Cl.MC/C57.1	Induction of MIP-1α production
Lipoteichoic acid	Peripheral blood monocytes	Induction of MIP-1α mRNA expression
IL-1 inhibitors	Mixed lymphocyte culture	Inhibition of MIP-1α production
Normal or IFN-γ activated, ICAM-1 expressing human endothelial cells	Peripheral blood cells	Co-culture induces MIP-1α production
Culture on plates coated with soluble ICAM-1	Peripheral blood cells	Induction of MIP-1α mRNA and protein expression
Hemazoin crystals	Peripheral blood monocytes	Induction of MIP-1α production
PMA and calcium ionophore	Peripheral blood monocytes	Induction of MIP-1α production
	Dendritic Langerhans' cells	Induction of MIP-1α production
CD40 ligand	Peripheral blood monocytes	Induction of MIP-1α production
	Dendritic Langerhans' cells	
PGE$_2$	Murine alveolar and peritoneal macrophages	Inhibition of LPS-induced MIP-1α production

Stimulus	Cell type	Response
Anti-CD3 antibody	Peripheral blood lymphocytes (CD8⁺CD45RO⁺ population)	Induction of MIP-1α production
	Murine T cells	Induction of MIP-1α mRNA in 30 min and MIP-1α protein in 6 h
Thrombin, ADP	Platelets	Release of MIP-1α from α-granules
Drugs		
Dexamethasone	LPS-stimulated murine alveolar and peritoneal macrophages	Inhibition of MIP-1α production

[a] When HL-60 cells were induced to differentiate into granulocytes by addition of retinoic acid, no MIP-1α mRNA was induced.
[b] Murine dendritic cell lines, which resemble Langerhans' cells, constitutively express MIP-1α mRNA.

Expression in disease

Disease	Tissue	Comment
Human disease		
RA	Synovial fluid macrophages and lining cells	Expression of MIP-1α mRNA and protein
RA	Serum	Increased expression in patients with high serum levels of rheumatoid factor
Sarcoidosis or interstitial pulmonary fibrosis	Bronchoalveolar lavage fluid	Detection of MIP-1α protein
HIV infection	Alveolar macrophages	Spontaneous expression of MIP-1α. MIP-1α protein secretion increases as viral infection develops and there is a correlation with CD8+ alveolitis. Infection of cells in culture does not induce MIP-1α expression.
Allergen challenge	Nasal secretions	Detection of MIP-1α protein. The level is significantly reduced following steroid treatment
Hypersensitivity[2]	Macrophages in the bronchoalveolar lavage	Expression of MIP-1α protein.
Multiple sclerosis	Cerebrospinal fluid	Elevated MIP-1α protein. The concentration correlates with leukocyte cell counts
Behçet's disease and HTLV-1 associated myelopathy	Cerebrospinal fluid	Significantly higher levels of MIP-1α
Microbial invasion of the amniotic cavity during gestation	Amniotic fluids	Increased MIP-1α protein levels in both term and preterm gestations which correlate with the white blood cell count
Acute *Staphyllococcus aureus* endo carditis	Endocardial samples containing neutrophils, macrophages, and fibroblasts	Cell-associated MIP-1α expression

Disease	Tissue	Comment
Malaria infection	Peripheral blood	Measurable MIP-1α which does not correlate with parasite count or hematocrit values
Aplastic anemia	Long term bone marrow cultures	Spontaneous production of elevated levels of MIP-1α

Animal models of disease

Bleomycin-induced lung injury	Alveolar macrophages and bronchial epithelial cells	Induction of time-dependent increase in MIP-1α which correlates with accumulation of granulocytes and mononuclear cells
Mice with collagen-induced arthritis	Joints	Detection of MIP-1α mRNA and protein. Administration of neutralizing antibodies delays the onset and reduces the severity of arthritis. Anti-IL-10 treatment enhances MIP-1α expression
Venous thrombosis in rats	Vein wall homogenates Rat model of cerebral ischemic brain tissue containing macrophages and endothelial cells	Detection of MIP-1α protein MIP-1α is found beginning 6 h post-ischemia following middle cerebral artery occlusion. Suppressed by steroid treatment
Murine model of experimental autoimmune encephalomyelitis	Spinal cord and brain parenchymal cells resembling astrocytes, encephalitogenic T cells	Expression of MIP-1α in correlation with the onset of histologic and clinical disease
Thioglycollate injection of mouse peritoneal cavity	Peritoneal exudate murine macrophages	Expression of MIP-1α mRNA 2–5 days after injection of thioglycollate
LPS injection of mice	Lung macrophages and the subendothelium	Expression of time dependent increase in MIP-1α mRNA and protein. Pretreatment with neutralizing antisera decreases the influx of neutrophils, lung permeability, and lung ICAM-1 mRNA and delays mortality. Lung mRNA is also reduced by neutralization of TNF prior to administration of LPS
Intratracheal inoculation of mice with *Klebsiella pneumoniae*	Lung	Increase in lung MIP-1α concentration Neutralization of IL-10 enhances expression
Mice with pulmonary granuloma formation following systemic *Schistosoma mansoni* egg infection	Macrophages within granulomas	Protein expression. Reduction in granuloma formation following antibody treatment.

Disease	Tissue	Comment
Intratracheal challenge of mice with *S. mansoni* egg antigen	Airway epithelial cells Alveolar macrophages Recruited mononuclear cells	Induction of MIP-1α mRNA and protein expression Administration of antibody to MIP-1α reduces eosinophil recruitment to the lungs
Intrapulmonary deposition of IgG immune complexes or LPS	Bronchoalveolar lavage fluid	Time dependent increase in MIP-1α mRNA Antibody administered at the time of challenge reduces neutrophil infiltration, lung vascular permeability and TNF-α levels in the lavage fluid.
Murine model of acute graft versus host disease	Spleen, gut and liver	Persistent expression of MIP-1α mRNA

[a]Human skin expresses mRNA which can be localized to the epidermal Langerhans' cell.
[b]Mouse skin expresses mRNA which can be localized to the epidermal Langerhans' cell.

In vivo studies

Animal studies

A single injection of human MIP-1α protein into canine dermis fails to induce a response.

Administration of purified murine MIP-1α protein induces a localized inflammatory response with neutrophil infiltration at 4 h when injected subcutaneously into the footpads of mice.

Multiple intraperitoneal injections of murine MIP-1α protein causes an increase in the total number of local macrophages and also increases neutrophils, mast cells, and eosinophils.

MIP-1α protein injected intravenously produces a monophasic fever of rapid onset which is not inhibited by administration of cyclooxygenase inhibitors.

MIP-1α injection by intracerebroventricular infusion into rats induces a long-lasting fever.

Murine MIP-1α protein injected into the anterior hypothalamic, preoptic area of rats induces fever.

Murine MIP-1α protein injected into the anterior hypothalamic, preoptic area of rats reduces food intake.

Murine MIP-1α protein injected into mice exhibits a reversible decrease in the cycling rates and in the absolute numbers of myeloid progenitor cells in the marrow and spleen[3].

Human MIP-1α protein injected into mice inhibits subpopulations of hematopoietic progenitors that are activated in myelosuppressed animals but has no effect on the long term reconstituting stem cells under conditions where it reduces all later progenitors.

Human MIP-1α protein injected into mice modifies the regenerating bone marrow population such that recovery from cytotoxic agents is initiated from the more primitive cells of the population.

Injection of MIP-1α transfected tumor cell lines demonstrates impaired growth in nude mice. Histologic examination of the injection site shows predominantly neutrophil infiltration.

Neutralizing anti-MIP-1α antibody administered to mice with bleomycin induced lung injury demonstrates reduced cell infiltration and fibrosis.

Homozygous MIP-1α mutant (−/−) mice are resistant to Coxsackie virus induced myocarditis seen in wild type animals. Influenza virus infected −/− mice have reduced pneumonitis and delayed viral clearance compared with wild type animals. The mice have no gross abnormalities in development or in the histology of any major organ and no hematopoietic abnormalities.

In vitro biological effects

T lymphocytes

Chemotaxis (0.01 nM–1 μM)
Inhibition of proliferation (5 nM)
Induction of adhesion (0.1–5 nM)
Shape change (10 nM)
Suppression of cytokine production (5 nM)
Mobilization of intracellular free calcium (50 nM)

Basophils

Histamine release (0.5–50 nM)
Mediator release (IL-3 sensitized) (1–100 nM)
Chemotaxis (0.3–30 nM)
Mobilization of intracellular free calcium (1–100 nM)
Inhibition of induced histamine release (5 nM)

Hematopoietic precursors

Inhibition of proliferation (0.1–100 nM) [4]
Inhibition of AML progenitor proliferation (5–200 nM)
Enhanced growth factor dependent progenitor proliferation (5–10 nM)
Up-regulation of adhesion (0.1–1 nM)
Increased phospholipid turnover (0.5–5 nM)
Elevation of cAMP (0.1–5 nM)

Monocytes

Chemotaxis (0.01–100 nM)
Chemotaxis – cell lines (0.1–100 nM)
Mobilization of intracellular free calcium (0.1–100 nM)
Mobilization of intracellular free calcium – cell lines (10–100 nM)

Expression of β_2 adhesion molecules (10 nM–1 μM)
Cytokine secretion (0.2–20 nM)
Enzyme release (3–100 nM)
Stimulation of proliferation (10 nM)
Enhanced antibody dependent killing (100–400 nM)
Release of arachidonic acid (5–10 nM)

Eosinophils

Actin polymerization (10 nM)
Mobilization of intracellular free calcium (10 nM)
Chemotaxis (1–100 nM)
Degranulation (10–100 nM)
Stimulation of oxidative metabolism (1–10 nM)

Neutrophils

Chemotaxis (1–100 nM)
Mobilization of intracellular free calcium (10–300 nM)
Oxidative burst (0.125–100 nM)

Mast cells

Histamine release (10 nM)

Natural killer cells

Chemotaxis (0.01–0.1 nM)
Chemotaxis of activated cells (0.1–1 nM)

Epidermal keratinocytes

Inhibition of proliferation (0.5–5 nM)

Astrocytes

Inhibition of proliferation (5–25 nM)

Spermatogonia

Increase in DNA synthesis (stage specific) (1–10 nM)
Inhibition of DNA synthesis (stage specific) (1–10 nM)

Dendritic cells

Chemotaxis (10 nM)
Mobilization of intracellular free calcium (10 nM)

Osteoclasts

Chemotaxis (10 nM)
Suppression of bone resorption (0.01–1 nM)

Receptor binding characteristics[5]

Binds to CCR1, CCR4, CCR5, and US28.

Cross-desensitization

In human basophils and eosinophils MIP-1α fails to desensitize a calcium flux response to MCP-1 and is only weakly active against RANTES. RANTES strongly desensitizes the response to MIP-1α in basophils and eosinophils. MCP-1 partially desensitizes subsequent eosinophil response to MIP-1α.

With human monocytes, MCP-3 desensitizes the calcium response to MIP-1α but exposure to MIP-1α only minimally reduces a subsequent MCP-3 response. Also in monocytes, MIP-1α desensitizes the calcium response to RANTES but not MCP-1 but RANTES does not desensitize monocytes to MIP-1α. In human monocytes, MIP-1α desensitizes enzyme release by RANTES but not MCP-1, MCP-2, or MCP-3. The MIP-1α response is desensitized by prior exposure to RANTES and MCP-3 and to a lesser extent by MCP-1 and MCP-2.

Calcium response in human monocytes is desensitized by prior exposure to MCP-1 or RANTES and prior exposure to MIP-1α desensitizes a response to RANTES but not MCP-1.

MIP-1α desensitizes THP-1 cells to a calcium response to MCP-1 or RANTES but MCP-1 or RANTES do not desensitize a subsequent MIP-1α response.

MIP-1α desensitizes dendritic cells to a calcium response to MCP-3 but not RANTES. Exposure of dendritic cells to MCP-3 or RANTES do not desensitize a calcium response to MIP-1α.

In HL-60 cells, MIP-1α and RANTES cross-desensitize.

Gene structure

Murine

A TATAA sequence is located at −32 bp. A DNase I hypersensitive site, which maps to the proximal promoter, is present in macrophage cells but is absent in cells which do not express basal levels of MIP-1α mRNA.

Human

The MIP-1α gene has three exons and two introns. The promoter contains a TATA box at –28.

Gene location

The human gene, located on the long arm of chromosome 17 (bands q11–q21), occurs as a single copy gene with a closely related non-allelic copy (referred to as 464.2) present in the genome of many individuals. A single copy gene (*Scya3*) is found on the distal portion of chromosome 11 near the *Hox-2* gene complex in mouse.

Protein structure

Human MIP-1α protein at physiological ionic strength can exist in monomeric and tetrameric forms which display distinct conformational properties[6]. The monomeric form is the active component on hematopoietic cells[7]. Mutation of leucine 25 in IL-8 to tyrosine creates a protein with monocyte chemo-attractant activity which displaces MIP-1α receptor binding. Biological activity on THP-1 cells can be inhibited by a peptide from the N-terminus of the CCR-1 receptor. Murine MIP-1α, unlike the human form, contains one potential *O*-glycosylation site. Murine MIP-1α exists in polymer form in solution[8]. The murine form also binds to heparin.

Amino acid sequence

Human

```
MQVSTAALAV LLCTMALCNQ FSASLAADTP TACCFSYTSR QIPQNFIADY
FETSSQCSKP GVIFLTKRSR QVCADPSEEW VQKYVSDLEL SA
```

Murine[9]

```
1   MKVSTTALAV LLCTMTLCNQ VFSYAPGADT PTACCFSYSR KIPRQFIVDY
    FETSSLCSQP GVIFLTKRNR QICADSKETW VQEYITDLEL NA
```

Murine

```
2   MKVSTTALAV LLCTMTLCNQ VLSYAPGADT PTACCFSYSR KIPRQFIVDY
    FETSSLCSQP GAIFLTKRNR QICADSKETW VQEYITDLEL NA
```

Two murine sequences with minor differences were reported.

Rat

```
1   MKVSTAALAV LLCTMALWNE VFSAPYGADT PTACCFSYGR QIPRKFIADY
    FETSSLCSQP GVIFLTKRNR QICADPKETW VQEYITELEL NA
```

Rat

```
2   MKVSTTALAV LLCTMALWNE VFSAPYGADT PTACCFSYGR QIPRKFIADY
    FETSSLWSQP GVIFLTKRNR QICADPKETW VQEYITELEL NA
```

Two rat sequences with minor differences were reported.

Database accession numbers

Species	SwissProt	EMBL/GenBank	PIR	Ref
Murine	P10855	M23447	A27596	10
	P14096	X12531	A30552	
		X53372	A32393	
		J04491	S04533	
		M73061	S11685	
Human	P10147	D00044	A24198	11
		M23452	A30574	
		M25315		
		X03754		
		X04018		
		M23178		
		D90144		
Rat		U22414		12

References
1 Wolpe, S.D. and Cerami, A. (1989) *FASEB J.* 3, 2565–2573.
2 Driscoll, K.E. (1994) *Exp. Lung Res.* 20, 473–490.
3 Lord, B. et al. (1994) *Int. J. Hematol.* 57, 197–206.
4 Graham, G.J. et al. (1990) *Nature* 344, 442–444.
5 Wang, J.M. et al. (1993) *J. Immunol.* 150, 3022–3029.
6 Patel, S.R. et al. (1994) *Biochemistry* 32, 5466–5471.
7 Lord, B.I. et al. (1995) *Blood* 85, 3412–3415.
8 Wolpe, S.D. et al. (1995) *J. Exp. Med.* 167, 570–581.
9 Davatelis, G. et al. (1988) *J. Exp. Med.* 167, 1939–1944.
10 Widmer, U. et al. (1993) *J. Immunol.* 150, 4996–5012.
11 Blum, S. et al. (1990) *DNA Cell Biol.* 9, 589–602.
12 Shanley, T.P. et al. (1995) *J. Immunol.* 154, 4793–4802.
13 Cocchi, F. et al. (1995) *Science* 270, 1811–1815.

MIP-1β   Macrophage inflammatory protein-1β

Alternate names

pAT744 gene product, Act-2, G-26, H-400 (murine), hSISγ (murine)

Family

β family (CC family)

Molecule

MIP-1β shares about 70% homology with MIP-1α[1]. The molecules are structurally very similar but there are significant differences in their functions. MIP-1β is inactive, unlike MIP-1α, in the activation of neutrophils. MIP-1α inhibits early hematopoitic progenitor growth, whereas MIP-1β potentiates it. In terms of HIV-suppressive effects, both MIP-1α and MIP-1β seem to have a synergistic effect.

Tissue sources

Monocytes[1], fibroblasts, T lymphocytes, B lymphocytes, neutrophils, smooth muscle cells, mast cells, tumor cell lines such as epithelial cell line A549.

Target cells

Monocytes, T lymphocytes, hematopoietic precursor cells and basophils

Physicochemical properties

Property	Human	Murine
pI (mature)	~9	~9
Signal	1–23	1–23
Amino acids		
Precursor	92	92
Mature	68	68
Disulfide bonds	a.a. 34–58	34–58
	a.a. 35–74	35–74
N-glycosylation sites	0	0
O-glycosylation sites	2	3
Molecular weight		
Precursor	10 199	10 215
SDS-PAGE	11 000–13 000	

Transcription factors

The human MIP-1β promoter contains three copies of the binding site for PU.1 (−132, −171, −263) and three consensus GRE sites (−393, −950, −1053). The murine promoter exhibits cell specific and LPS-inducible expression, and contains a TATA box (−26), four PU.1 sites and two GRE sites. There is an essential cAMP response element although the promoter is not cAMP responsive. There is a c-*jun* binding site which is distinct from the phorbol ester response element.

Regulation of expression

Monocytes

MIP-1β mRNA and protein are induced in human peripheral blood monocytes by addition of either lipopolysaccharide or IL-7[1].

Induction of MIP-1β in human cells is inhibited by addition of IL-4 but appears to be unresponsive to addition of steroids.

Purified or chemically synthesized hemazoin crystals induce MIP-1β secretion from human cells.

The U937 human histiocytic lymphoma cell line expresses MIP-1β mRNA following phorbol ester stimulation.

Murine cell lines express MIP-1β mRNA and protein in response to addition of serum, LPS, or cycloheximide.

Bone marrow derived murine macrophage expression of MIP-1β mRNA is inhibited by TGF-β.

Fibroblasts

Human synovial fibroblast cultures from RA patients produce MIP-1β mRNA in response to addition of IL-1α.

Human pulmonary fibroblasts constitutively produce MIP-1β protein that is not enhanced by addition of LPS, IL-1β, or TNF-α.

T lymphocytes

Human peripheral blood lymphocytes stimulated with PHA express MIP-1β mRNA.

Selected T cell lines express MIP-1β mRNA constitutively or when activated by mitogens.

Murine and human Th1 and Th2 cell lines express MIP-1β mRNA following Con A stimulation.

Antigen stimulated naive murine splenic T cells express MIP-1β mRNA.

B lymphocytes

Human peripheral blood lymphocytes stimulated with *Staphylococcus aureus* extract express MIP-1β mRNA.

Neutrophils

Human neutrophils treated with IFN-γ express MIP-1β mRNA.

Human granulocytes stimulated with lipopolysaccharide express MIP-1β mRNA and protein and this induction can be suppressed by addition of IL-10.

Smooth muscle cells

Pulmonary smooth muscle cells express MIP-1β protein in response to addition of TNF-α, IL-1β, IL-4, IL-10, or IFN-γ.

Mast cells

The human mast cell line HMC-1 expresses MIP-1β mRNA upon activation by phorbol esters.

MIP-1β mRNA is detectable in the murine mast cell lines Cl.MC/9, Cl.MC/9.2 and Cl.MC/C57.1 after FcϵRI engagement with antigen or Con A stimulation.

Tumor cell lines

The epithelial cell line A549 constitutively produces MIP-1β protein that is not enhanced by addition of LPS, IL-1β, or TNF-α.

Expression in disease

Human studies

MIP-1β mRNA can be identified in synovial mononuclear cells of rheumatoid arthritis patients.

Elevated levels of MIP-1β mRNA are found in circulating peripheral blood and synovial fluid T lymphocytes from rheumatoid arthritis patients. By immunohistochemistry, expression is located in sites of lymphocyte infiltration and proliferation and expression is seen in non-T and T cell populations.

Animal studies

MIP-1β mRNA is expressed during the course of a murine model of experimental autoimmune encephalomyelitis in correlation with the onset of histologic and clinical disease. MIP-1β mRNA is expressed in encephalitogenic T cells.

In a murine model of *Escherichia coli* pyelonephritis, MIP-1β mRNA was expressed in kidneys after bacterial challenge.

In vivo effects

Animal studies

Murine MIP-1β injected into the anterior hypothalamic preoptic area of rats induces fever[2].

Murine MIP-1β injected into the anterior hypothalamic preoptic area of rats reduces food intake.

In vitro biological effects

T lymphocytes

Chemotaxis (0.01 nM–1 μM)[3]
Induction of adhesion (0.01–5 nM)
Mobilization of intracellular free calcium (50 nM)
Actin polymerization[3]

Monocytes

Chemotaxis (0.01 –100 nM)
Enzyme release (30–100 nM)

Hematopoietic precursors

Inhibition of proliferation (0.5–12.5 nM)
Inhibition of MIP-1α suppression of proliferation (10–20 nM)

Basophils

Histamine release (10–100 nM)
Inhibition of induced histamine release (5 nM)

Intracellular signaling

Causes transient change in intracellular calcium in monocytes but not in neutrophils.

Does not cause GTPase activation in NK cells.

Receptor binding characteristics

Binds to CCR1, CCR5, and US28.

Cross-desensitization

Does not cross-desensitize monocytes for subsequent stimulation with MCP-1.

Does not cross-desensitize neutrophils for subsequent stimulation with IL-8. MIP-1β and MIP-1α cross-desensitize each other for calcium mobilization in monocytes.

Gene structure

Human

The gene is organized into three exons with two introns. A consensus TATAA sequence is found at –26. There is an additional nonallelic copy (referred to as 744.2) in the genome of many individuals.

Murine[4]

The gene is organized into three exons and two introns.

Gene location

The human MIP-1β gene is located on the long arm of chromosome 17 (bands q11–q21). The murine gene is located on the distal portion of chromosome 11 near the *Hox-2* gene complex.

Protein structure

High resolution NMR solution structure of the human protein demonstrates a dimer stabilized by hydrogen bonding and hydrophobic interactions. The main secondary structure elements are comprised of a triple stranded antiparallel β-sheet (residues 26–31, 39–44, and 48–52) arranged in a Greek key, on top of which lies an α-helix (residues 57–68)[5].

Amino acid sequence

Human[7]

```
MKLCVTVLSL LMLVAAFCSP ALSAPMGSDP PTACCFSYTA RKLPRNFVVD
YYETSSLCSQ PAVVFQTKRS KQVCADPSES WVQEYVYDLE LN
```

Murine[6]

```
MKLCVSALSL LLLVAAFCAP GFSAPMGSDP PTSCCFSYTS RQLHRSFVMD
YYETSSLCSK PAVVFLTKRG RQICANPSEP WVTEYMSDLE LN
```

Rabbit

```
MKLGVTVLSV ALLVAALCPP ALSAPMGSDP PTACCFSYTL RKLPRHFVID
YFETTSLCSQ PAVVFQTKKG RQVCANPSES WVQEYVDDLE LN
```

Chicken

```
MKVSVAALAV LLIAICYQTS AAPVGSDPPT SCCFTYISRQ LPFSFVADYY
ETNSQCPHAG VVFITRKGRE VCANPENDWV QDYMNKLELN
```

Database accession numbers

Species	SwissProt	EMBL/GenBank	PIR	Ref
Human	P22617	M57503		9
Murine	P14097	M23503	C30552	8
		M35590	JL0088	
Rabbit	P46632	D17402		10
Chicken		L34553		11

References

1 Ziegler, S.F. et al. (1994) *J. Immunol.* 147, 2234–2239.
2 Myers, R.D. et al. (1993) *Neurochem. Res.* 18, 667–673.
3 Adams, D.H. et al. (1994) *Proc. Nat. Acad. Sci. USA* 91, 7144–7148.
4 Proffitt, J. et al. (1995) *Gene* 152, 173–179.
5 Lodi, P.J. et al. (1994) *Science* 263, 1762–1767.
6 Sherry, B. et al. (1988) *J. Exp. Med.* 168, 2251–2259.
7 Lipes, M.A. et al. (1988) *Proc. Nat. Acad. Sci.* USA 85, 9704–9708.
8 Widmer, U. et al. (1993) *J. Immunol.* 150, 4996–5012.
9 Hillier, L. et al. (1995) Direct GenBank submission from the Washu-Merck EST Project.
10 Mori, S. Direct GenBank submission.
11 Petrenko, O. and Enrietto, P.J. (1994) Direct GenBank submission.

RANTES

 Regulated on activation, normal T cell expressed and secreted

Alternate names

TY-5 (murine)

Family

β family (CC family)

Molecule

RANTES was originally identified by molecular cloning as a transcript expressed in T cells but not B cells. It is the only β-chemokine known to be present in platelets, and has potent chemotactic and activating properties for basophils, eosinophils, and NK cells. It has HIV-suppressive effect and synergizes with MIP-1α and MIP-1β in the suppression of HIV[12].

Tissue sources

T lymphocytes, epithelial cells, monocytes, fibroblasts, mesangial cells, platelets, eosinophils, T cell lines (Hut 78), NK cell lines, HEL cell line, HL 60 cells, human mast cell line HMC-1, human lung carcinoma cell line CALU-3, and human melanoma cells.

Target cells

Monocytes, T lymphocytes, basophils, eosinophils, NK cells, dendritic cells, and mast cells.

Physicochemical properties

Property	Human	Murine
pI (mature)	9.5	~9
Signal	1–23	1–23
Amino acids		
Precursor	91	91
Mature	68	68
Disulfide bonds	a.a. 33–57	33–57
	a.a. 34–73	34–73
N-glycosylation sites	0	0
Molecular weight		
Predicted	10 075	10 071
SDS-PAGE	8000	8000
ESP-MS	7863	

Transcription factors

Potential regulatory motifs in the human RANTES promoter include NF-κB (–30), AP-1 (–327, –345, –354), Ets-1 (–372), myb (–310), and CD28RE (–579).

Regulatory motifs in the murine RANTES promoter include a PU.1 box, an NF-κB, and an IRF-1 response element.

Regulation of expression

Stimulus	Cell type	Response
Cytokines		
IL-1 and TNF-α	Human lung epithelial cell line A549	Stimulation of RANTES mRNA expression, is inhibited by steroids
	Murine renal tubular epithelial cell line	Elevation of RANTES expression
	Synovial fibroblasts	Induction of RANTES expression
	Mouse mesangial cell line MMC	Induction of RANTES expression
TNF-α, IFN-γ	Human bronchial epithelial cell line BEAS-2B	Stimulation of RANTES mRNA expression; steroids inhibit expression of both mRNA and protein
IL-4	Human synovial fibroblasts	Inhibition of RANTES mRNA expression induced by IL-1α and β and TNF-α
Others		
PHA[1]	Normal human peripheral T cells	Induction of RANTES mRNA and protein expression three to seven days after activation
PMA	Human mast cell line HMC-1	Expression of RANTES mRNA
PMA or anti-CD3[2] stimulation	Peripheral blood lymphocytes (CD8+CD45RO+ population)	Induction of RANTES mRNA and protein expression within one day of activation
Thrombin	Platelets	Release of active protein from α granules

[a] IL-2 dependent antigen specific cytotoxic and helper human T cell clones express RANTES mRNA and protein but not established T cell tumor lines[3]
[b] The human T cell line HA1.7 expresses high levels of RANTES mRNA constitutively and levels are not changed by either tolerizing or activation treatments. Despite the high levels of mRNA, no protein is detected in the supernatant
[c] Nephritogenic antigen specific murine T cell clones express RANTES mRNA constitutively.
[d] High levels of RANTES mRNA are constitutively expressed in mouse macrophage cell lines RAW 264.7 and PU5-1.8.
[e] Purified human eosinophils express RANTES mRNA and protein.

Expression in disease

Human studies

Pelvic fluid concentrations of RANTES are elevated in women with endometriosis and the levels correlate with the severity of the disease.

RANTES mRNA is detected in synovial tissue samples from patients with rheumatoid arthritis.

Elevated levels of RANTES mRNA are found in circulating peripheral blood and synovial fluid T lymphocytes from RA patients. By immunohisto-chemistry, RANTES expression is located in sites of lymphocyte infiltration and proliferation and is limited to T cells.

RANTES mRNA and protein are detected in infiltrating mononuclear cells and tubular epithelium during acute renal allograft rejection. RANTES protein, but not RANTES mRNA, is also expressed on inflamed endothelium[4].

RANTES mRNA is expressed in allergen challenged skin sites in atopic subjects.

RANTES protein is detected in nasal secretions following allergen challenge. The level is significantly reduced following steroid treatment[5].

RANTES protein is expressed in the epithelium of nasal polyps, a lesion characterized by an eosinophil rich infiltrate.

Bronchial biopsies from asthmatics express RANTES protein in the bronchial epithelium and this expression is down-regulated by treatment with inhaled steroids.

RANTES protein is detectable in the diluted tears of patients with allergic conjunctivitis.

Animal studies

RANTES mRNA is expressed during the course of a murine model of experimental autoimmune encephalomyelitis in correlation with the onset of histologic and clinical disease. RANTES mRNA is expressed in encephalitogenic T cells.

In a rat model of chronic rejection of renal allografts, RANTES mRNA is expressed between the early and late phases of chronic rejection.

A time dependent increase in RANTES mRNA and protein is seen in the lungs of mice after systemic exposure to lipopolysaccharide. RANTES protein is immunolocalized to alveolar epithelial cells. Pretreatment with neutralizing antibody blocks lung infiltration of macrophages but not of neutrophils or lymphocytes.

RANTES mRNA is expressed in a model of anti-GMB glomerulonephritis in mice.

In vivo studies

Animal studies

A single intradermal injection of human RANTES protein into canine dermis results in eosinophil and macrophage rich inflammatory sites within 4 h.

Administration of human RANTES protein by intracerebroventricular infusion into rats decreases short term food intake.

In vitro biological effects

Monocytes

Chemotaxis (0.01–100 nM)[6]
Chemotaxis (cell lines) (6–300 nM)[7]
Mobilization of intracellular free calcium (50–100 nM)
Mobilization of intracellular free calcium – cell lines (10–100 nM)
Expression of β_2 adhesion molecules (10 nM–1 μM)
Enzyme release (3–100 nM)
Release of arachidonic acid (5–10 nM)

T lymphocytes

Chemotaxis (0.01 nM–1 μM)
Induction of cell polarization (0.01–100 nM)
Mobilization of intracellular free calcium (1 nM–1 μM)
Induction of IL-2 receptor (1 μM)
Induction of cytokine production (1 μM)
Enhanced cell proliferation (1 μM)
Induction of adhesion (0.01–1 nM)
Stimulation of PI$_3$-kinase (1–100 nM)

Basophils

Mediator release (nonsensitized) (0.5–100 nM)
Mediator release (IL-3 sensitized) (10–100 nM)
Inhibition of induced histamine release (0.1–100 nM)
Chemotaxis (0.3–100 nM)
Mobilization of intracellular free calcium (10–100 nM)

Eosinophils

Mobilization of intracellular free calcium (1–100 nM)
Chemotaxis (1–100 nM)
Actin polymerization (1–100 nM)
Adhesion to ICAM-1 (10 nM)
Stimulation of oxidative metabolism (1–100 nM)
Degranulation (10–100 nM)
IL-8 release after GMCSF priming

Natural killer cells

Chemotaxis (0.01–0.1 nM)
Chemotaxis, activated cells (0.01 nM)

Dendritic cells

Chemotaxis (10 nM)
Mobilization of intracellular free calcium (10 nM)

Mast cells

Chemotaxis (0.01–100 nM)

Receptor binding characteristics

Binds to CCR1, CCR3, CCR5, DARC, and US28.

Cross-desensitization

On basophils and eosinophils, cross-desensitization of calcium flux is seen with RANTES and MIP-1α. However, while RANTES desensitizes subsequent response to MIP-1α the reverse gives little to no effect in some studies while another study finds equivalent desensitization. MCP-1 does not affect the response to RANTES. MCP-2 and RANTES cross-desensitize in a basophil calcium response but while RANTES can desensitize a calcium response to MCP-2 in eosinophils, MCP-2 has no effect on a subsequent RANTES response.

With THP-1 cells, prior incubation with RANTES does not inhibit response to MCP-1 or MIP-1α but MCP-1 or MIP-1α desensitize the response to RANTES. MIP-1α desensitizes THP-1 cells to calcium response to RANTES but RANTES does not desensitize a subsequent MIP-1α response.

With human monocytes, MCP-3 desensitizes the calcium response to RANTES but exposure to RANTES only minimally reduces a subsequent MCP-3 response. In human monocytes, RANTES desensitizes enzyme release by MIP-1α and to a lesser extent by MCP-2 and MCP-3. The RANTES response is desensitized by prior exposure to MCP-2, MCP-3, and MIP-1α and to a lesser extent by MCP-1.

The calcium response in human monocytes is desensitized by prior exposure to MCP-1 or MIP-1α and prior exposure to RANTES desensitizes a response to MIP-1α but not MCP-1.

In HL-60 cells, MIP-1α and RANTES cross-desensitize.

Gene structure

Human

Murine

Human

The RANTES gene locus spans approximately 7.1 kb and is composed of three exons of 133, 112, and 1075 bases and two introns of approximately 1.4 and 4.4 kb. The promoter contains a TATAAA box at −12 and a CCAAT box at −70.

Murine[8]

The RANTES gene is organized into three exons and two introns. The transcriptional start site is located 27 bp downstream of a TATAA box.

Gene location

The human RANTES gene is found in a single copy on chromosome 17 (q11.2–12). The murine genome contains one gene, *Scya5*, located on chromosome 11.

Protein structure

Three-dimensional solution structure analysis by NMR spectroscopy demonstrates a solution dimer at low pH with dimerization occurring between the N-terminal regions. Each monomer consists of a three-stranded antiparallel β-sheet (residues 26–30, 38–43, 48–51) in a Greek key motif with a C-terminal helix (56–65) packed across the sheet and two short N-terminal β-strands. The dimer dissociates into a monomer with a K_d of 35 μM[9].

Amino acid sequence

Human[10]

```
MKVSAARLAV ILIATALCAP ASASPYSSDT TPCCFAYIAR PLPRAHIKEY
FYTSGKCSNP AVVFVTRKNR QVCANPEKKW VREYINSLEM S
```

Murine

```
1  MKISAAALTI ILTAAALCAP APASPYGSDT TPCCFAYLSL ALPRAHVKEY
   FYTSSKCSNL AVVFVTRRNR QVCANPEKKW VQEYINYLEM S
```

Murine[8]

```
2  MKISAAALTI ILTAAALCTP APASPYGSDT TPCCFAYLSL ELPRAHVKEY
   FYTSSKCSNL AVVFVTRRNR QVCANPEKKW VQEYINYLEM S
```

Two murine sequences with minor differences were reported.

Database accession numbers

Species	SwissProt	EMBL/GenBank	PIR	Ref
Human	P13501	M21121	A28815	3
Murine	P30882	M77747		11
		S37648		
		U02298		8

References

1. Nelson, P.J. et al. (1993) *J. Immunol.* 151, 2601–2612.
2. Schall, T.J. (1991) *Cytokine* 3, 165–183.
3. Schall, T.J. et al. (1988) *J. Immunol.* 141, 1018–1025.
4. Pattison, J.M. et al. (1995) *Clin. Immunother.* 4, 1–8.
5. Zhang, L. et al. (1994) *Clin. Exp.Allergy* 24, 899–904.
6. Schall, T.J. et al. (1990) *Nature* 347, 669–671.
7. Wang, J.M. et al. (1993) *J. Exp. Med.* 177, 699–705.
8. Danoff, T.M. et al. (1994) *J. Immunol.* 152, 1182–1189.
9. Chung, C.-w. et al. (1995) *Biochemistry* 34, 9307–9314.
10. Wiedermann, C. et al. (1993) *Curr. Biol.* 3, 735–739.
11. Neilson, E.G. et al. (1992) *Kidney Int.* 41, 220–225.
12. Cocchi, F. et al. (1995) *Science* 270, 1811–1815.

Alternate names

TCA3 (murine), P500 (murine), SISε (murine)

Family

β family (CC family)

Molecule

I-309 is secreted by activated T cells and is a monocyte chemoattractant. The primary structure of I-309 contains two features that are not present in other β-chemokines, i.e., two additional cysteine residues and a single N-linked glycosylation site located in the middle of the β-strand that forms the proposed dimer interface. I-309 is also shown to be a monomer at high concentrations.

Tissue sources

T lymphocytes, human mast cells, HMC-1, murine mast cell lines Cl.MC/9, Cl.MC/9.2, and Cl.MC/C57.1

Target cells

Monocytes, macrophages, neutrophils, basophils, and microglial cells

Physicochemical properties

Property	Human	Murine
pI (mature)	~9	~9
Signal	1–23	1–23
Amino acids		
Precursor	96	92 or 95
Mature	73	69 or 72
Disulfide bonds	a.a. 33–57	33–57
	a.a. 34–73	34–74
N-Glycosylation sites	1 (a.a. 52)	1
O-Glycosylation sites	1	1
Molecular weight		
Precursor	10992	10276
SDS-PAGE	12000	
N-Glycosylated	15–16000	14–16000

Transcription factors

There is an NF-κB-like sequence in the I-309 promoter at –201 and a PU.1 site at –287.

Regulation of expression

Human cells express I-309 mRNA in response to addition of PHA and expression can be inhibited by addition of steroids[1].

I-309 mRNA is expressed in the human T cell line IDP2 and in phorbol ester and PHA stimulated Jurkat cells[2].

I-309 mRNA is expressed in antigen-, IL-2-, or Con A-treated murine T cells. Cyclosporin A blocks this expression[3].

The human mast cell line HMC-1 expresses I-309 mRNA and protein upon activation by phorbol esters.

I-309 mRNA is detectable in the murine mast cell lines Cl.MC/9, Cl.MC/9.2, and Cl.MC/C57.1 after FcεRI engagement with antigen or Con A stimulation.

In vivo studies

Animal studies

I-309 mRNA is expressed during the course of a murine model of experimental autoimmune encephalomyelitis in correlation with the onset of histologic and clinical disease. mRNA is expressed in encephalitogenic T cells[4].

Administration of purified murine I-309 protein induces a localized inflammatory response with neutrophil and macrophage infiltration after 2 h when injected intraperitoneally. Increased numbers of neutrophils and monocytes are also observed in the peripheral blood[5].

I-309 transfected tumor cell lines demonstrate impaired growth in both normal and immunodeficient mice and I-309 expression specifically augments tumor immunogenicity[6].

Injection of murine I-309 protein into the footpads of mice induces swelling, apparent by 1 h and peaking at 2 h and remaining elevated out to 19 h. At 2 h, there is a striking accumulation of neutrophils at the injection site[7].

In vitro biological effects

Monocytes/Macrophages

Chemotaxis (3 nM–1 μM)[8]
Mobilization of intracellular free calcium (10–20 nM)

Neutrophils

Chemotaxis (1–100 nM)

Basophils

Histamine release (10 nM)

Microglial cells

Chemotaxis (0.1–10 nM)

Gene structure

Human

The human gene is composed of three exons (76, 112, and 100 bp of coding sequence in each, sequentially). There is no apparent CCAAT box but there is an unconventional TATA box (AATAA) at position –29.

Murine

The gene is divided into three exons.

Gene location

The human gene, designated *Scya1*, is located on chromosome 17. The murine gene is located on the distal portion of mouse chromosome 11 near the *Hox-2* gene complex.

Amino acid sequence

Human[10]

```
MQIITTALVC LLLAGMWPED VDSKSMQVPF SRCCFSFAEQ EIPLRAILCY
RNTSSICSNE GLIFKLKRGK EACALDTVGW VQRHRKMLRH CPSKRK
```

Murine

```
MKPTAMALMC LLLAAVWIQD VDSKSMLTVS NSCCLNTLKK ELPLKFIQCY
RKMGSSCPDP PAVVFRLNKG RESCASTNKT WVQNHLKKVN PC
```

Murine (alternative splice product)[9]

```
MKPTAMALMC LLLAAVWIQD VDSKSMLTVS NSCCLNTLKK ELPLKFIQCY
RKMGSSCPDP PAVVVRSSGV PGLTEAEKTV TDSSE
```

Database accession numbers

Species	SwissProt	EMBL/GenBank	PIR	Ref
Human	P22362	M57502 M57506	A45817 A37236	2
Murine	P10146	M17957	S24236	11

References

[1] Selvan, R.S. et al. (1994) *J. Biol. Chem.* 269, 13893–13898.
[2] Miller, M.D. et al. (1990) *J. Immunol.* 145, 2737–2744.
[3] Burd, P.R. et al. (1987) *J. Immunol.* 139, 3126–3131.
[4] Godiska, R.D. et al. (1995) *J. Neuroimmunol.* 58, 167–176.
[5] Luo, Y. et al. (1994) *J. Immunol.* 153, 4616–4624.
[6] Laning, J. et al. (1994) *J. Immunol.* 153, 4625–4635.
[7] Wilson, S.D. et al. (1990) *J. Immunol.* 145, 2745–2750.
[8] Miller, M.D. and Krangel, M.S. (1992) *Proc. Natl. Acad. Sci. USA* 89, 2950–2954.
[9] Brown, K.D. et al. (1989) *J. Immunol.* 142, 679–687.
[10] Miller, M.D. et al. (1989) *J. Immunol.* 143, 2907–2916.
[11] Direct GenBank submission.

Family

β family (CC family)

Molecule

Eotaxin was identified by microsequencing of HPLC purified proteins from bronchoalveolar lavage fluid taken from ovalbumin sensitized guinea pigs[1]. Eotaxin exhibits 53% homology with human MCP-1, 44% with guinea pig MCP-1, 31% with human MIP-1α, and 26% with human RANTES. The eotaxin gene is widely expressed in normal mice and is strongly induced in cultured endothelial cells in response to IFN-γ. Eotaxin is also induced locally in response to the transplantation of IL-4-secreting tumour cells, indicating that it likely contributes to the eosinophil recruitment and antitumour effect of IL-4. The human homolog of eotaxin has recently been reported[2].

Tissue sources

Murine endothelial cells, alveolar macrophages, lung, intestine, stomach, heart, thymus, spleen, liver, testes, kidney.

Target cells

Eosinophils.

Physicochemical properties

Property	Human	Guinea pig	Murine
Signal	?	?	23
Amino acids			
Precursor	97	73	97
	?	?	?
Disulfide bonds	a.a. 32–57	8–33	32–57
	a.a. 33–73	9–49	33–73
Glycosylation sites	?	1 (a.a. 70)	?
Molecular weight			
Calculated	8150	8330	8200
	8380		
	8810		
	9030		

Regulation of expression

Eotaxin mRNA is undetectable in a murine endothelial cell line but treatment with IFN-γ rapidly induces its expression.

Expression in disease

Eotaxin is constitutively expressed in the lungs of guinea pigs. Lower levels are detectable in the intestine, stomach, heart, thymus, spleen, liver, testes, and kidney[1].

Eotaxin protein and mRNA are increased in the lungs of sensitized guinea pigs within thirty minutes following challenge with aerosolized antigen in association with eosinophil infiltration[3].

Eotaxin is constitutively expressed in murine skin, thymus, lymph node, mammary gland and skeletal muscle with lower levels also seen in heart, stomach, tongue and lung.

Eotaxin mRNA is expressed at the site of implantation of IL-4 expressing tumour cells in mice.

In vivo studies

Animal studies

Eotaxin protein injected into guinea pigs intradermally induces eosinophil but not neutrophil accumulation.

Exposure of guinea pigs to aerosolized Eotaxin protein induces eosinophil but not neutrophil accumulation in the bronchoalveolar fluid after twenty hours[3].

In vitro biological effects

Eosinophils

Chemotaxis
Mobilization of intracellular free calcium (0.5–5 nM)
Aggregation (10–40 nM)

Gene location

Murine[5]

A single copy gene designated *Scya11* maps to chromosome 11 between *D11Mit* markers 7 and 36.

Amino acid sequence

Human

```
MKVSAALLWL LLIAAAFSPQ GLAGPASVPT TCCFNLANRK IPLQRLESYR
RITSGKCPQK AVIFKTKLAK DICADPKKKW VQDSMKYLDQ KSPTPKP
```

Guinea pig[6]

```
MKVSTAFLCL LLTVSAPSAQ VLAHPGIPSA CCFRVTNKKI SFQRLKSYKI
ITSSKCPQTA IVFEIKPDKM ICADPKKKWV QDAKKYLDQI SQTTKP
```

Murine

MOSSTALLFL LLTVTSFTSQ VLAHPGSIPT SCCFIMTSKK IPNTLLKSYK
RITNNRCTLK AIVFKTRLGK EICADPKKKW VQDATKHLDQ KLQTPKP

Database accession numbers

Species	SwissProt	EMBL/GenBank	Ref
Human		U34780	2
Guinea pig	P80325	X77603	1
Mouse		U26426	7

References
[1] Rothenberg, M.E. et al. (1955). *J. Exp. Med.* 181, 1211–1216.
[2] Ponath, P.D. et al. (1996) *J. Clin. Invest.* 97, 604–612.
[3] Jose, P.J. et al. (1994) *J. Exp. Med.* 179, 881–887.
[4] Griffiths-Johnson, D.A. et al. (1994) *Biochem. Biophys. Res. Comm.* 197, 1167–1172.
[5] Rothenberg, M.E. et al. (1995) *Proc. Natl. Acad. Sci. USA* 92, 8960–8964.
[6] Jose, P.J. et al. (1994) *Biochem. Biophys. Res. Comm.* 205, 788–794.
[7] Jia, G.Q. et al. (1996) *Immunity* 4, 1–14.

Family

β family (CC family)

Molecule

C10 has only been reported in the mouse. No human homolog is known. The biological significance of C10 is not known.

Tissue sources

Bone marrow cells, myeloid cell lines, macrophages, T lymphocytes.

Physicochemical properties

Property	Murine
pI (mature)	9.4
Signal	1–21
Amino acids	
Precursor	116
Mature	95
Disulfide bonds	a.a. 50–73
	a.a. 51–89
Glycosylation sites	1 (a.a. 29)
Molecular weight	
Precursor	12 984
SDS-PAGE	13 000

Regulation of expression

C10 is expressed by GMCSF-treated murine bone marrow cells[1].

C10 mRNA is expressed in two immature murine myeloid cell lines.

C10 mRNA is expressed by the P388D1 murine cell line as well as by murine macrophages produced from bone marrow cultures.

C10 is expressed by IL-3, IL-4 or granulocyte-macrophage colony stimulating factor treated murine bone marrow or resident peritoneal cells[2].

C10 mRNA is expressed by an IL-2 dependent murine cell line.

C10 mRNA is induced during induction of differentiation of 32DcI3 cells by GCSF.

Gene structure

Murine

In contrast to most members of the β-chemokine family, the gene is organized into a four-exon and three-intron structure.

Gene location

Murine[3]

The single gene is located on the distal portion of chromosome 11 closely linked to *Scya2* and is designated *Scya6*.

Amino acid sequence

Murine

```
MRNSKTAISF FILVAVLGSQ AGLIQEMEKE DRRYNPPIIH QGFQDTSSDC
CFSYATQIPC KRFIYYFPTS GGCIKPGIIF ISRRGTQVCA DPSDRRVQRC
LSTLKQGPRS GNKVIA
```

Database accession numbers

Species	SwissProt	EMBL/GenBank	Ref
Murine	P27784	M58004	1
		L11237	

References

[1] Orlofsky, A. et al. (1991) *Cell Regulation* 2, 403–412.
[2] Orlofsky, A. et al. (1994) *J. Immunol.* 152, 5084–5091.
[3] Berger, M.S. et al. (1993) *DNA and Cell Biol.* 12, 839–847.

Family

β family (CC family)

Molecule

HCC-1 was originally isolated from the hemofiltrate of patients with chronic renal failure. Mature HCC-1 has 46% sequence identity with MIP-1α and MIP-1β, and 29–37% with the other human β-chemokines. Unlike, other β-chemokines, HCC-1 is expressed constitutively in several normal tissues (spleen, liver, skeletal and heart muscle, gut, and bone marrow), and is present at high concentrations (1–80 nM) in plasma.

Tissue sources

Plasma, spleen, liver, skeletal and heart muscle, gut, and bone marrow.

Target cells

Monocytes and hematopoietic progenitor cells.

Expression in disease

Human studies

HCC-1 protein is present in the plasma of normal individuals. HCC-1 mRNA is expressed in several normal tissues[1].

HCC-1 protein is isolated from the hemofiltrate of patients with chronic renal failure.

In vitro biological effects

Monocytes

Mobilization of intracellular free calcium (1 μM)

Hematopoietic progenitor cells

Enhanced proliferation (100 nM)

Gene structure

Human

Amino acid sequence

Human

```
MKISVAAIPF FLLITIALGT KTESSSRGPY HPSECCFTYT TYKIPRQRIM
DYYETNSQCS KPGIVFITKR GHSVCTNPSD KWVQDYIKDM KEN
```

Database accession numbers

Species	SwissProt	EMBL/GenBank	Ref
Human		Z49270	1

Reference

[1] Schulz-Knappe, P. et al. (1996) *J. Exp. Med.* 183, 295–299.

Alternate names

Ltn. A homologous protein named ATAC gene product is also reported.

Family

γ family (C family)

Molecule

Human lymphotactin (Ltn) shows similarity to some members of the β-chemokine family but is missing the first and third cysteine residues that are characteristic of the α- and β-chemokines[1]. Lymphotactin is chemotactic for lymphocytes but not for monocytes[2], a characteristic that makes it unique among chemokines. In addition, calcium flux desensitization studies indicate that Ltn uses a unique receptor[2]. A cDNA clone, designated ATAC, isolated from a collection of human T cell activation genes encodes a protein 73.8% identical to murine lymphotactin[3].

Tissue sources

Thymocytes, activated T cells.

Target cells

T lymphocytes.

Physicochemical properties

The structural features predict the cleavage and secretion of a mature Ltn protein of approximately 10 kD from the 12.52 kD precursor.

Regulation of expression

Activated immature thymocytes (pro-T cells) express Ltn mRNA. Further analysis of T cell subsets demonstrates Ltn mRNA presence in double negative thymocytes and activated CD8+ but not CD4+ or CD8+CD4+ T cells[1].

Ltn mRNA, of approximately 0.9 kb, is exclusively expressed in activated CD8+ T cells.

Induction of the Ltn gene requires stimulation by both PMA and Ca^{2+} ionophore A23187 ('two-signal gene') and is fully abrogated by the immunosuppressive agent cyclosporin A. Upon stimulation, Ltn mRNA is detectable within 30 min of stimulation, and maximal expression is seen after 4 h.

In vitro biological effects

T lymphocytes

Chemotaxis (0.1–100 nM)
Mobilization of intracellular free calcium (1 μM)

Bone marrow cells

Chemotaxis (0.1–100 nM)

Gene location

Murine

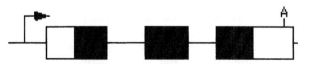

The human ATAC gene is located on chromosome 1q23[3].

Amino acid sequence

Human

```
MRLLILALLG ICSLTAYIVE GVGSEVSDKR TCVSLTTQRL PVSRIKTYTI
TEGSLRAVIF ITKRGLKVCA DPQATWVRDV VRSMDRKSNT RNNMIQTKPT
GTQQSTNTAV TLTG
```

Murine

```
MRLLLLTFLG VCCLTPWVVE GVGTEVLEES SCVNLQTQRL PVQKIKTYII
WEGAMRAVIF VTKRGLKICA DPEAKWVLAA IKTVDGRAST RKNMAETVPG
TGAQRSTSTA ITLTG
```

Rat

```
MRLLLLTFLG VCCFAAWVVE GVGTEVLQES ICVSLRTQRL PVQKIKTYTI
KEGAMRAVIF VTKRGLRICA DPQAKWVKTA IKTVDGRASA SKSKAETIPT
QAQRSASTAV TLTG
```

Database accession numbers

Species	EMBL/GenBank	Ref
Mouse	U15607	1
Human	U23772	2
Rat	U23377	4

References
[1] Kelner, G.S. et al. (1995) *Science* 266, 1395–1399.
[2] Kennedy, J. et al. (1995) *J. Immunol.* 155, 203–209.
[3] Muller, S. et al. (1995) *Eur. J. Immunol.* 25, 1744–1748.
[4] Jones, M.L. Direct GenBank submission.

SCM-1   Single cysteine motif-1

Family

γ family (C family)

Molecule

SCM-1 is encoded by a cDNA clone isolated from human peripheral blood mononuclear cells stimulated with PHA[1]. SCM-1, which is significantly related to the α- and β-chemokines, has only the second and fourth of the four cysteines conserved in these proteins. SCM-1 is 60.5% identical to lymphotactin. SCM-1 and lymphotactin may represent the human and murine prototypes of a γ (C) chemokine family.

Tissue sources

Human T cells and spleen[1].

Physicochemical properties

Property	Human
Signal	?
Amino acids	
Precursor	114
	?
Disulfide bonds	a.a. 32–69
Glycosylation sites	?
Molecular weight	12508

Regulation of expression

SCM-1 is strongly induced in human PBMC and Jurkat T cells by PHA stimulation.

Gene location

The *SCM-1* gene is distinctly mapped to human chromosome 1[1].

Amino acid sequence

Human

```
MRLLILALLG ICSLTAYIVE GVGSEVSDKR TCVSLTTQRL PVSRIKTYTI
TEGSLRAVIF ITKRGLKVCA DPQATWVRDV VRSMDRKSNT RNNMIQTKPT
GTQQSTNTAV TLTG
```

Database accession numbers

Species	GenBank	Ref
Human	D43768	1

Reference
[1] Yoshida, T. *et al.* (1995) *FEBS Lett.* 360, 155–159.

THE
CHEMOKINE
RECEPTORS

Alpha (CXC) Chemokine Receptors

Alternate names

High affinity IL-8 receptor, class I IL-8R

Family

G-protein coupled receptor family, chemokine receptor branch of rhodopsin family

Homologs

IL-8RB, equine herpesvirus protein[1]

Tissue sources

Human neutrophils[2]

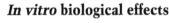

In vitro biological effects

- COS-7 cells expressing IL-8RA show an influx of calcium following IL-8 addition[2].
- HEK 293 cells expressing IL-8RA show an influx of calcium following IL-8 addition, $K_d = 3.03 \pm 1.01$[3].
- Calcium influx and chemotaxis stimulated by treatment with IL-8 in Jurkat cells stably expressing IL-8A is inhibited by PT[4].
- Jurkat cells stably transfected with IL-8RA chemotax in response to IL-8, GRO and NAP-2[5].
- There is a transient rise in intracellular calcium in response to IL-8 in cells expressing the rabbit IL-8 receptor.

Animal models

Murine IL-8 receptor knockout mice showed splenomegaly, enlarged cervical lymph nodes, extramedullary myelopoiesis and compromised neutophil acute migration[6].

Ligands and ligand binding studies

Affinities of IL-8, MGSA, and NAP-2 for IL-8RA are:
- IL-8 (2 nM) >>> MGSA/GRO (450 nM) = NAP-2[7].
- COS-7 cells expressing human IL-8RA bind ^{125}I-IL-8 with high affinity[2].
- COS-7 cells expressing rabbit IL-8R bind ^{125}I-IL-8 with high affinity[8].
- Jurkat cells stably expressing IL-8RA bind IL-8, ($K_d = 3-5$ nM)[4].
- Jurkat cells stably expressing IL-8RA bind IL-8 ($K_d = 1-4$ nM), GRO-α and NAP-2 ($K_d = 200-500$ nM)[5].

Expression pattern

Hematopoietic cells

IL-8RA mRNA is found in neutrophils[2,6], monocytes, basophils; freshly isolated T cells[9].

Cell lines

HL-60

Regulation of expression

- Using flow cytometry with receptor-specific antibodies, 119 nM IL-8 was shown to down-regulate IL-8RA surface expression within 5 min; expression is restored within 90 min[10].
- G-CSF and LPS regulate IL-8RA mRNA.

Gene structure

Human

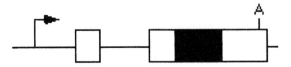

- Chromosomal location: 2q33–q36[21].
- Exon/intron organization: There are two exons with the coding region contained in a single exon, and an Alu repeat in the intron[11].
- Promoter: GC-rich 5'-flanking region with silencer elements at –841 and –280, and an Alu upstream from the promoter[11].

Protein structure

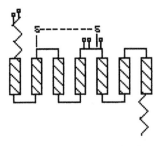

- 350 amino acids; calculated MW = 39 806; cross-linking data compute size to be 58–67 kD[12].
- Putative glycosylation sites: Asn-3, Asn-16, Asn-181, 182, Asn-193.
- Putative disulfide bridge: Cys-110–Cys-187.
- Mutagenesis: N-terminal domain confers ligand-binding specificity[13]; Asp-11, Glu-275, Arg-280 important for IL-8 binding[14]; Arg-199, Arg-203, and

Asp-265 are involved in IL-8 binding and signal transduction[3]; rabbit receptor with N-terminal point mutations to the corresponding human sequence (H13Y, T15K) binds human IL-8 with high affinity[15].

In vitro signal transduction

IL-8 stimulation of Jurkat cells stably expressing IL-8RA results in the phosphorylation of p42/p44 MAP kinase[4].

Amino acid sequence

Human IL-8RA

```
1   MSNITDPQMW DFDDLNFTGM PPADEDYSPC MLETETLNKY VVIIAYALVF
51  LLSLLGNSLV MLVILYSRVG RSVTDVYLLN LALADLLFAL TLPIWAASKV
101 NGWIFGTFLC KVVSLLKEVN FYSGILLLAC ISVDRYLAIV HATRTLTQKR
151 HLVKFVCLGC WGLSMNLSLP FFLFRQAYHP NNSSPVCYEV LGNDTAKWRM
201 VLRILPHTFG FIVPLFVMLF CYGFTLRTLF KAHMGQKHRA MRVIFAVVLI
251 FLLCWLPYNL VLLADTLMRT QVIQESCERR NNIGRALDAT EILGFLHSCL
301 NPIIYAFIGQ NFRHGFLKIL AMHGLVSKEF LARHRVTSYT SSSVNVSSNL
```

Murine IL-8R

```
1   MGEFKVDKFN IEDFFSGDLD IFNYSSGMPS ILPDAVPCHS ENLEINSYAV
51  VVIYVLVTLL SLVGNSLVML VILYNRSTCS VTDVYLLNLA IADLFFALTL
101 PVWAASKVNG WTFGSTLCKI FSYVKEVTFY SSVLLLACIS MDRYLAIVHA
151 TSTLIQKRHL VKFVCIAMWL LSVILALPIL ILRNPVKVNL STLVCYEDVG
201 NNTSRLRVVL RILPQTFGFL VPLLIMLFCY GFTLRTLFKA HMGQKHRAMR
251 VIFAVVLVFL LCWLPYNLVL FTDTLMRTKL IKETCERRDD IDKALNATEI
301 LGFLHSCLNP IIYAFIGQKF RHGLLKIMAT YGLVSKEFLA KEGRPSFVSS
351 SSANTSTTL
```

Rabbit IL-8R

```
1   MEVNVWNMTD LWTWFEDEFA NATGMPPVEK DYSPCLVVTQ TLNKYVVVVI
51  YALVFLLSLL GNSLVMLVIL YSRSNRSVTD VYLLNLAMAP AFCPDHAYLG
101 RLQGKRLDFR TPLCKVVSLV KEVNFYSGIL LLACISVDRY LAIVQSTRTL
151 TQKRHLVKFI CLGIWALSLI LSLPFFLFRQ VFSPNNSSPV CYEDLGHNTA
201 KWCMVLRILP HTFGFILPLL VMLFCYGFTL RTLFQAHMGQ KHRAMRVIFA
251 VVLIFLLCWL PYNLVLLADT LMRTHVIQET CQRRNELDRA LDATEILGFL
301 HSCLNPIIYA FIGQNFRNGF LKMLAARGLI SKEFLTRHRV TSYTSSSTNV
351 PSNL
```

Database accession numbers

Source	PIR	SwissProt	EMBL/GenBank	Ref
Human	A39445	P25024	X65858, M68932, L19592	2
Mouse			U15208, L13239	16
			D17630	17

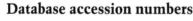

Source	PIR	SwissProt	EMBL/GenBank	Ref
Rabbit	A23669	P21109	M58021, J05705	18
	A46483		L24445	19
Gorilla			X91110	20
Chimpanzee			X91109	20
Orangutan			X91111	20

References

1. Telford, E.A. et al. (1995) *J. Mol. Biol.* 249, 520–528.
2. Holmes, W.E. et al. (1991) *Science* 253, 1278–1280.
3. Leong, S.R. et al. (1994) . *J. Biol. Chem.* 269, 19343–19348.
4. Jones, S.A. (1995) *FEBS Lett.* 364, 211–214.
5. Loetscher, M. et al. (1994) *FEBS Lett.* 341, 187–192.
6. Cacalano, G. et al. (1994) *Science* 265, 682–684.
7. Lee, J. et al. (1992) *J. Biol. Chem.* 267, 16283–16287.
8. Beckman, M.P. et al. (1991) *Biochem. Biophys. Res. Comm.* 179, 784–789.
9. Xu, L. et al. (1995) *J. Leukoc. Biol.* 57, 335–342.
10. Chuntharapai, A. and Kim, K.J. (1995) *J. Immunol.* 155, 2587–2594.
11. Sprenger, H. et al. (1994) *J. Immunol.* 153, 2524–2532.
12. Begemer, J. et al. (1989) *J. Biol. Chem.* 264, 17409–17417.
13. LaRosa, G.J. et al. (1992) *J. Biol. Chem.* 267, 25402–25406.
14. Hebert, C.A. et al. (1993) *J. Biol. Chem.* 268, 18549–18553.
15. Schraufstatter, I.U. et al. (1995) *J. Biol. Chem.* 270, 10428–10431.
16. Lee, J. et al. (1995) *J. Immunol.* 155, 2158–2164.
17. Harada, A. et al. (1993) *Int. Immunol.* 5, 681–690.
18. Thomas, K.M. et al. (1990) *J. Biol. Chem.* 265, 20061–20064.
19. Prado, G.N. et al. (1994) *J. Biol. Chem.* 269, 12391–12394.
20. Alvarez, V. (1996) *Immunogen.* 43, 261–267.
21. Mollereau, C. et al. (1993) *Genomics* 16, 248–251.

Alternate names

Low affinity IL-8R, class II IL-8 receptor[1], CXCR2, F3R (rabbit)

Homologs

IL-8RAP pseudogene; equine herpesvirus protein[2], IL-8RA

Tissue sources

Dibutyryl cAMP-treated HL-60 cells[1]

In vitro biological effects

- *Xenopus* oocytes microinjected with IL-8RB mobilize calcium ($EC_{50} = 20$ nM)[1].
- HEK 293 cells expressing IL-8RB chemotax in response to 10–100 ng/ml of IL-8[3].
- Jurkat cells stably transfected with IL-8RB bind IL-8, $K_d = 3$–5 nM[4].
- Jurkat cells stably expressing IL-8RB chemotax in response to IL-8, GRO-α and NAP-2[5].
- GTP-γS activation of 3ASubE cells expressing IL-8RB is PT-sensitive[6].
- IL-8 binding to COS cells expressing IL-8RB activates a phosphatidylinositol-calcium second messenger system in a PT-sensitive manner[7].
- Binding of IL-8 to IL-8RB enhances CD11b expression[8].
- Binding of IL-8 to IL-8RB leads to receptor phosphorylation and degradation[6].
- Undifferentiated HL-60 cells transfected with IL-8RB showed calcium mobilization and actin polymerization in response to IL-8[9].

Ligands and ligand binding studies

- α-chemokines IL-8 (2 nM) >> MGSA/GRO (2 nM) > NAP-2.
- KC (PDGF-induced in mouse fibroblasts) is the mouse homolog of Gro-α.
- *Xenopus* oocytes microinjected with IL-8RB bind ^{125}I-IL-8[1].
- HEK 293 cells expressing IL-8RB bind IL-8, $K_d = 0.1$–0.2 nM[3].
- HEK 293 cells expressing IL-8RB bind IL-8, $K_d = 1.3$ nM[9].
- Jurkat cells stably expressing IL-8RB bind IL-8 with high affinity[5].
- 3ASubE cells expressing IL-8RB bind MGSA, $K_d = 0.3$–0.7 nM[6].
- COS cells expressing hIL-8R type A and B bind IL-8 and IL-8, GRO and NAP-2 respectively, with high affinity[10].
- CHO cells expressing mIL-8R do not bind IL-8 in the nanomolar range[11].
- COS cells expressing mIL-8R transiently bind both KC and MIP-2 with high affinity[12].
- Membranes of COS-7 cells expressing rabbit F3R bind both human IL-8 and rabbit IL-8[13].

Expression pattern

Cells

IL-8RB mRNA is found in neutrophils[14], monocytes, basophils, and T cells[15], primary keratinocytes[6].

Cell lines

IL-8RB mRNA is found in HL-60, THP-1[1]; human placental cell line 3ASubE P-3[6]; melanoma cell lines[16]; Hs294T[17].

Expression pattern in human disease

IL-8RB is expressed in the lesional skin of patients with psoriasis[18].

Regulation of expression

- IL-8 and MGSA down-regulate IL-8RB surface expression on human neutrophils within 5 min of treatment, as assessed by flow cytometry[19].
- hIL-8RB on T lymphocytes is down-regulated by culturing with or without anti-CD3 for 12 h or more at 37°C[15].
- G-CSF up-regulates IL-8RB mRNA expression[20].

Gene structure

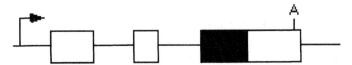

- Genomic organization: 12 kb gene with the coding region entirely contained in the third exon. There are two separate reports of the structure of the gene's untranslated regions: 11 exons of 5'-UT[21] or 2 exons of 5'UT and a 1.4 kb 3'UT[20].
- Transcripts: 7 mRNAs (alternative splicing of the 5'UT[21]; 3.0 kb mRNA[1].
- Chromosomal location: 2q34–q35[22], maps near *Ity-Lsh-Bcg* locus in the mouse[23].
- Promoter: TATA 47 bp upstream from exon 1[21], TATA 20 bp upstream of exon 1 along with GC-rich region with three SP-1 and two AP-2 binding sites, G-CSF-responsive element at –118 and silencers between –779 and –118[20].

Protein structure

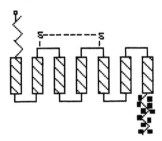

- 360 amino acids; calculated MW = 40 123 D, 42 kD in 3ASubE cells expressing IL-8RB[6].

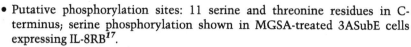

- Putative phosphorylation sites: 11 serine and threonine residues in C-terminus; serine phosphorylation shown in MGSA-treated 3ASubE cells expressing IL-8RB[17].
- Putative glycosylation site: Asn-17.
- Putative disulfide bond: Cys-114–Cys-191.
- Mutagenesis: N-terminal domain confers ligand-binding specificity[7,24]; residues 317–324 are involved in signaling and chemotaxis[3].

In vitro signal transduction

- Tyrosine phosphorylation of the crk-associated substrate, p130/cas, after MGSA stimulation of 3ASubE cells over-expressing IL-8RB[25].
- Phosphorylation of p42/p44 MAP kinase by IL-8 or GRO in Jurkat cells stably transfected with IL-8RB[4].

Amino acid sequence

Human IL-8R

```
1   MESDSFEDFW KGEDLSNYSY SSTLPPFLLD AAPCEPESLE INKYFVVIIY
51  ALVFLLSLLG NSLVMLVILY SRVGRSVTDV YLLNLALADL LFALTLPIWA
101 ASKVNGWIFG TFLCKVVSLL KEVNFYSGIL LLACISVDRY LAIVHATRTL
151 TQKRYLVKFI CLSIWGLSLL LALPVLLFRR TVYSSNVSPA CYEDMGNNTA
201 NWRMLLRILP QSFGFIVPLL IMLFCYGFTL RTLFKAHMGQ KHRAMRVIFA
251 VVLIFLLCWL PYNLVLLADT LMRTQVIQET CERRNHIDRA LDATEILGIL
301 HSCLNPLIYA FIGQKFRHGL LKILAIHGLI SKDSLPKDSR PSFVGSSSGH
351 TSTTL
```

Rabbit

```
1   MQEFTWENYS YEDFFGDFSN YSYSTDLPPT LLDSAPCRSE SLETNSYVVL
51  ITYILVFLLS LLGNSLVMLV ILYSRSTCSV TDVYLLNLAI ADLLFATTLP
101 IWAASKVHGW TFGTPLCKVV SLVKEVNFYS GILLLACISV DRYLAIVHAT
151 RTMIQKRHLV KFICLSMWGV SLILSLPILL FRNAIFPPNS SPVCYEDMGN
201 STAKWRMVLR ILPQTFGFIL PLLVMLFCYV FTLRTLFQAH MGQKHRAMRV
251 IFAVVLIFLL CWLPYNLVLL TDTLMRTHVI QETCERRNDI DRALDATEIL
301 GFLHSCLNPI IYAFIGQKFR YGLLKILAAH GLISKEFLAK ESRPSFVASS
351 SGNTSTTL
```

Mouse

```
1   MGEFKVDKFN IEDFFSGDLD IFNYSSGMPS ILPDAVPCHS ENLEINSYAV
51  VVIYVLVTLL SLVGNSLVML VILYNRSTCS VTDVYLLNLA IADLFFALTL
101 PVWAASKVNG WTFGSTLCKI FSYVKEVTFY SSVLLLACIS MDRYLAIVHA
151 TSTLIQKRHL VKFVCIAMWL LSVILALPIL ILRNPVKVNL STLVCYEDVG
201 NNTSRLRVVL RILPQTFGFL VPLLIMLFCY GFTLRTLFKA HMGQKHRAMR
251 VIFAVVLVFL LCWLPYNLVL FTDTLMRTKL IKETCERRDD IDKALNATEI
301 LGFLHSCLNP IIYAFIGQKF RHGLLKIMAT YGLVSKEFLA KEGRPSFVSS
351 SSANTSTTL
```

Database accession numbers

Source	PIR	SwissProt	EMBL/GenBank	gi#	Ref
Human	A39446, A53611	P25025	M73969, M94582	124358	1
Rabbit	A53752	P35344	L24445	547719	26
Mouse	A53677,48921	P35343	U31207	950175	27
			L13239		10
			D17630	493671	28

References

[1] Murphy, P.M. and Tiffany, H.L. (1991) *Science* 253, 1280–1283.

[2] Telford, E.A. et al. (1995) *J. Mol. Biol.* 249, 520–528.

[3] BenBaruch, A. et al. (1995) *J. Biol. Chem.* 270, 9121–9128.

[4] Jones, S.A. et al. (1995) *FEBS Lett.* 364, 211–214.

[5] Loetscher, M. et al. (1994) *FEBS Lett.* 341, 187–192.

[6] Mueller, S.G. et al. (1995) *J. Biol. Chem.* 269, 1973–1980.

[7] LaRosa, G.J. et al. (1992) *J. Biol. Chem.* 267, 25402–25406.

[8] L'Heureux, G.P. et al. (1995) *Blood* 85, 522–531.

[9] Schraffstatter, I.U. (1993) *J. Immunol.* 151, 6418–6428.

[10] Cerretti, D.P. et al. (1993) *Mol. Immunol.* 30, 359–367.

[11] Suzuki, H. et al. (1994) *J. Biol. Chem.* 269, 18263–18266.

[12] Bozic, C.R. et al. (1994) *J. Biol. Chem.* 269, 29355–29358.

[13] Thomas, K.M. et al. (1994) *J. Immunol.* 152, 2496–2500.

[14] Lee J. et al. (1992) *J. Biol. Chem.* 267, 16283–16287.

[15] Xu, L. et al. (1995) *J. Leukoc. Biol.* 57, 335–342.

[16] Moser, B. et al. (1993) *Biochem. J.* 294, 285–292.

[17] Mueller, S.G. et al. (1995) *J. Biol. Chem.* 270, 10439–10448.

[18] Lemster, B.H. et al. (1995) *Clin. Exp. Immunol.* 99, 148–154.

[19] Chuntharapai, A. and Kim, K.J. (1995) *J. Immunol.* 155, 2587–2594.

[20] Sprenger, H. et al. (1994) *J. Biol. Chem* 269, 11065–11072.

[21] Ahuja, S.K. et al. (1994) *J. Biol. Chem.* 269, 26381–26389.

[22] Mollereau, C. et al. (1993). *Genomics* 16, 248–251.

[23] Cerretti, D.P. et al. (1993) *Genomics* 18, 410–413.

[24] Gayle, R.B. et al. (1993) *J. Biol. Chem.* 268, 7283–7289.

[25] Schraw, W and Richmond, A. (1995) *Biochemistry* 34, 13760–13767.

[26] Prado, G.N. et al. (1994) *J. Biol. Chem.* 269, 12391–12394.

[27] Lee, J. et al. (1995) *J. Immunol.* 155, 2158–2164.

[28] Harada, A. et al. (1993) *Int. Immunol.* 5, 681–690.

[29] Heinrich, J.N. and Bravo, R. (1995) *J. Biol. Chem.* 270, 4987–4989.

Beta (CC) Chemokine Receptors

CC CKR-1 — CC chemokine receptor 1

Alternate names

MIP-1α receptor, HM145[1], LD78 receptor[1], MIP-1α/RANTES receptor[2], RANTES receptor[3], CCR-1[4]

Family

G-protein coupled receptor family, chemokine receptor branch of rhodopsin family

Homologs

- 80% amino acid identity between human and murine MIP-1R[5]
- Equine herpesvirus 2 protein[6]
- Murine MIP-1α RL1 and 2[5,17]

Tissue sources

HL-60[2]; human monocytes[1,4]

In vitro biological effects

- PT-sensitive G-protein coupling[11].
- Inhibition of adenylate cyclase by MIP-1α ($IC_{50} = 110$ pM) and RANTES ($IC_{50} = 140$ pM) in HEK 293 cells expressing CC CKR-1[12].
- HEK 293 cells expressing CC CKR-1 show an intracellular calcium flux in response to MIP-1α and RANTES[4].
- HEK 293 cells expressing CC CKR-1 chemotax in response to MCP-3, MIP-1α, and RANTES[13].
- *Xenopus* oocytes injected with CC CKR-1 RNA show a calcium influx in response to MIP-1α (100–5000 nM) and RANTES ($EC_{50} = 50$ nM), but not MIP-1β or MCP-1[2].
- Actin polymerization[14].
- Homologous and heterologous desensitization of cells expressing CC CKR-1 with MIP-1α and RANTES[4].

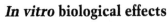

Ligands and ligand binding studies

- MIP-1α, MIP-1β, MCP-1, RANTES, MCP-3.
- $K_d > 100$ nM for MIP-1β and MCP-1[4].
- MIP-1α binds to COS-7 cells expressing CC CKR-1[2].
- HEK 293 cells expressing CC CKR-1 bind ^{125}I-MIP-1α with a $K_d = 5.1 \pm 0.3$ and ^{125}I-RANTES with $K_d = 7.6 \pm 1.5$ nM. There is heterologous displacement of MIP-1α by RANTES, MIP-1β and MCP-1, but MIP-1β and MCP-1 cause little calcium flux and may be acting as partial agonists[4].
- HEK 293 cells expressing CC CKR-1 bind ^{125}I-MCP-3, MIP-1α, and RANTES[11].

- COS cells transfected with murine CC CKR-1 bind human MIP-1α ($K_d = 3.4$ nM), RANTES ($K_d = 4.2$ nM) and murine MIP-1 ($EC_{50} = 8.9$ nM)[13].

Expression pattern
Hematopoietic cells
Neutrophils, monocytes, eosinophils, B cells[2], T cells

Tissues
Placenta, lung, and liver[8,9]

Cell lines
THP-1, U937, HL-60[2,4]

Expression in human disease
CC CKR-1 RNA is detected in peripheral blood and synovial fluid of RA patients, not osteoarthritis patients[10].

Gene structure

- There is a 102 bp 5' untranslated region followed by a 1065 bp ORF and 998 bp of 3' untranslated[2].
- Chromosomal localization: human chromosome 3p21[2], mouse chromosome 9[15].

Protein structure

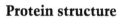
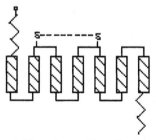

- 355 amino acids; calculated MW = 41 172 D
- Putative phosphorylation sites: none determined
- Putative glycosylation sites: Asn-5
- Putative disulfide bonds Cys-106–Cys-183

Amino acid sequences

Human CC CKR-1R

```
1   METPNTTEDY DTTTEFDYGD ATPCQKVNER AFGAQLLPPL YSLVFVIGLV
51  GNILVVLVLV QYKRLKNMTS IYLLNLAISD LLFLFTLPFW IDYKLKDDWV
101 FGDAMCKILS GFYYTGLYSE IFFIILLTID RYLAIVHAVF ALRARTVTFG
151 VITSIIIWAL AILASMPGLY FSKTQWEFTH HTCSLHFPHE SLREWKLFQA
201 LKLNLFGLVL PLLVMIICYT GIIKILLRRP NEKKSKAVRL IFVIMIIFFL
251 FWTPYNLTIL ISVFQDFLFT HECEQSRHLD LAVQVTEVIA YTHCCVNPVI
301 YAFVGERFRK YLRQLFHRRV AVHLVKWLPF LSVDRLERVS STSPSTGEHE
351 LSAGF
```

Mouse CC CKR-1R

```
1   MEISDFTEAY PTTTEFDYGD STPCQKTAVR AFGAGLLPPL YSLVFIIGVV
51  GNVLVILVLM QHRRLQSMTS IYLFNLAVSD LVFLFTLPFW IDYKLKDDWI
101 FGDAMCKLLS GFYYLGLYSE IFFIILLTID RYLAIVHAVF ALRARTVTLG
151 IITSIITWAL AILASMPALY FFKAQWEFTH RTCSPHFPYK SLKQWKRFQA
201 LKLNLLGLIL PLLVMIICYA GIIRILLRRP SEKKVKAVRL IFAITLLFFL
251 LWTPYNLSVF VSAFQDVLFT NQCEQSKHLD LAMQVTEVIA YTHCCVNPII
301 YVFVGERFWK YLRQLFQRHV AIPLAKWLPF LSVDQLERTS SISPSTGEHE
351 LSAGF
```

Rat CC CKR-1R

```
1   METPNTTEDY DTTTEFDYGD ATPCQKVNER AFGAQLLPPL YSLVFVIGLV
51  GNILVVLVLV QYKRLKNMTS IYLLNLAISD LLFLFTLPFW IDYKLKDDWV
101 FGDAMCKILS GFYYTGLYSE IFFIILLTID RYLAIVHAVF ALRARTVTFG
151 VITSIIIWAL AILASMPGLY FSKTQWEFTH HTCSLHFPHE SLREWKLFQA
201 LKLNLFGLVL PLLVMIICYT GIIKILLRRP NEKKSKAVRL IFVIMIIFFL
251 FWTPYNLTIL ISVFQDFLFT HECEQSRHLD LAVQVTEVIA YTHCCVNPVI
301 YAFVGERFRK YLRQLFHRRV AVHLVKWLPF LSVDRLERVS STSPSTGEHE
351 LSAGF
```

Database accession numbers

	SwissProt	GenBank	gi number	Ref
Human	P32246	L09230		2
Mouse		U28404		5
		U29678		13
Rat			741970	14

References

[1] Nomura, H. et al. (1993) *Int. Immunol.* 5, 1239–1249.

[2] Gao, J.L. et al. (1993) *J. Exp. Med.* 177, 1421–1427.

[3] Van Riper, G. et al. (1993) *J. Exp. Med.* 177, 851–856.

[4] Neote, K. et al. (1993) *Cell* 72, 415–425.

[5] Gao, J.-L. and Murphy, P.M. (1995) *J. Biol. Chem.* 270, 17494–17501.

[6] Telford, E.A. et al. (1995) *J. Mol. Biol.* 249, 520–528.

[7] Post, T.W. et al. (1995) *J. Immunol.* 155, 5299–5305.

[8] Combadiere, C. et al. (1995) *J. Biol. Chem.* 270, 29671–29675.

[9] Ahuja, S.K. and Murphy, P.M. (1993) *J. Biol. Chem.* 268, 20691–20694.

[10] Snowden. N. et al. (1994) *Lancet* 343, 547–548.

[11] Van Riper, G. et al. (1994) *J. Immunol.* 154, 4055–4061.

[12] Adams, D. H. et al. (1994) *Proc. Natl. Acad. Sci. USA* 91, 7144–7148.

[13] Ben-Baruch, A. et al. (1995) *J. Biol. Chem.* 270, 22123–22128.

[14] Harrison, J.K. et al. (1994) *Neurosci. Lett.* 169, 85–89

[15] Kozak, C.A. et al. (1995) *Genomics* 29, 294–296

CC CKR-2   CC chemokine receptor 2

Alternate names

MCP-1 receptor, CCR2 A and B[1]

Family

G-protein coupled receptor family, chemokine receptor branch of rhodopsin family

Tissue sources

MonoMac 6 (human monocytic leukemia cell line)[1], THP-1 cells[2]

In vitro biological effects

- Calcium mobilization is seen in *Xenopus* oocytes microinjected with CC CKR-2 A or B ($EC_{50} = 10-15$ nM)[1] and in HEK 293 cells expressing CC CKR-2B in response to MCP-1[3].
- HEK 293 cells expressing CC CKR-2B show an inhibition of adenylate cyclase in response to MCP-1 with an $IC_{50} = 90$ pM but no PI turnover was detected[3].
- PT-sensitive calcium mobilization and inhibition of adenylate cyclase in response to MCP-1 was observed in HEK 293 cells expressing CC CKR-2 stably[3].
- HEK 293 cells expressing CC CKR-2 B show homologous desensitization to MCP-1[1].

Ligands and ligand binding studies

- MCP-1, MCP-3.
- HEK-293 cells expressing CC CKR-2 bound ^{125}I-MCP-1 with $K_d = 260$ pM[3].
- HEK 293 cells expressing CC CKR-2 bound MCP-1 and MCP-3[4].

Expression pattern

Hematopoietic cells

Monocytes[5]

Tissues

Kidney, heart, bone marrow[2], lung, liver, pancreas[5]

Cell lines

MonoMac 6, THP-1 cells[1]

Gene structure

- 3.5 kb mRNAs.
- Alternative splicing of C-terminal tail, creating the A and B forms[1].

Protein structure

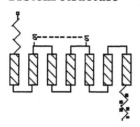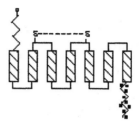

- 360 amino acids (B form) or 374 amino acids (A form).
- Putative phosphorylation sites: nine serine/threonine sites in the B form and five in the A form.

Amino acid sequence

Human MCP-1RA

```
1    MLSTSRSRFI RNTNESGEEV TTFFDYDYGA PCHKFDVKQI GAQLLPPLYS
51   LVFIFGFVGN MLVVLILINC KKLKCLTDIY LLNLAISDLL FLITLPLWAH
101  SAANEWVFGN AMCKLFTGLY HIGYFGGIFF IILLTIDRYL AIVHAVFALK
151  ARTVTFGVVT SVITWLVAVF ASVPGIIFTK CQKEDSVYVC GPYFPRGWNN
201  FHTIMRNILG LVLPLLIMVI CYSGILKTLL RCRNEKKRHR AVRVIFTIMI
251  VYFLFWTPYN IVILLNTFQE FFGLSNCEST SQLDQATQVT ETLGMTHCCI
301  NPIIYAFVGE KFRSLFHIAL GCRIAPLQKP VCGGPGVRPG KNVKVTTQGL
351  LDGRGKGKSI GRAPEASLQD KEGA
```

Human MCP-1RB

```
1    MLSTSRSRFI RNTNESGEEV TTFFDYDYGA PCHKFDVKQI GAQLLPPLYS
51   LVFIFGFVGN MLVVLILINC KKLKCLTDIY LLNLAISDLL FLITLPLWAH
101  SAANEWVFGN AMCKLFTGLY HIGYFGGIFF IILLTIDRYL AIVHAVFALK
151  ARTVTFGVVT SVITWLVAVF ASVPGIIFTK CQKEDSVYVC GPYFPRGWNN
201  FHTIMRNILG LVLPLLIMVI CYSGILKTLL RCRNEKKRHR AVRVIFTIMI
251  VYFLFWTPYN IVILLNTFQE FFGLSNCEST SQLDQATQVT ETLGMTHCCI
301  NPIIYAFVGE KFRRYLSVFF RKHITKRFCK QCPVFYRETV DGVTSTNTPS
351  TGEQEVSAGL
```

Database accession numbers

Source	GenBank	Ref
Human	U03882, U03905	1
	JC2443 (B form)	2
	D29984	

References

1. Charo, I.F. et al. (1994) *Proc. Natl. Acad. Sci.* 91, 2752–2756.
2. Yamagami, S. et al. (1994) *Biochem. Biophys. Res. Commun.* 202, 1156–1162.
3. Myers, S.J. et al. (1995) *J. Biol. Chem.* 270, 5786–5792.
4. Franci, C. et al. (1995) *J. Immunol.* 1545, 6511–6517.
5. Combadiere, C. et al. (1995) *J. Biol. Chem.* 270, 16491–16494.

CC CKR-3 CC chemokine receptor 3

Alternate names

Eosinophil chemokine receptor[1], RANTES receptor, CCR3, eotaxin receptor[2]

Family

G-protein coupled receptor family, chemokine receptor branch of rhodopsin family

Tissue sources

Human monocyte cDNA library[3]

In vitro biological effects

- Calcium influx[3].
- Decrease in cAMP accumulation[3].
- HEK 293 cells stably expressing CC CKR3 showed calcium influx in response to MIP-1α, MIP-1β and RANTES, not MCP-3[3].

Protein structure

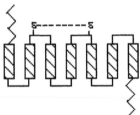

355 amino acids; calculated MW = 41 002 D.

Amino acid sequence

```
1    MTTSLDTVET FGTTSYYDDV GLLCEKADTR ALMAQFVPPL YSLVFTVGLL
51   GNVVVVMILI KYRRLRIMTN IYLLNLAISD LLFLVTLPFW IHYVRGHNWV
101  FGHGMCNLLS GFYHTGLYSE IFFIILLTID RYLAIVHAVF ALRARTVTFG
151  VITSIVTWGL AVLAALPEFI FYETEELFEE TLCSALYPED TVYSWRHFHT
201  LRMTIFCLVL PLLVMAICYT GIIKTLLRCP SKKKYKAIRL IFVIMAVFFI
251  FWTPYNVAIL LSSYQSILFG NDCERSKHLD LVMLVTEVIA YSHCCMNPVI
301  YAFVGERFRK YLRHFFHRHL LMHLGRYIPF LPSEKLERTS SVSPSTAEPE
351  LSIVF
```

Database accession numbers

	EMBL/GenBank	gi#	Ref
Human	HSU28694	881570	1
	U49727		2

References

1. Combadiere, C. et al. (1995) *J. Biol. Chem.* 270, 16491–16494.
2. Ponath, P.D. et al. (1996) *J. Exp. Med.* 183(6), 2437–2448.
3. Nomura, H. et al. (1993) *Int. Immunol.* 5, 1239–1249.

CC CKR-4
CC chemokine receptor 4

Alternate names

K5-5[1], CCR4, HS CCCR3[1]

Tissue sources

Immature basophilic cell line KU-812[1]

In vitro biological effects

MIP-1α-, RANTES-, and MCP-1-treatment resulted in stimulation of a Ca^{2+} activated chloride channel in *Xenopus* oocytes microinjected with CC CKR-4 mRNA[1].

Ligands and ligand binding studies

MIP-1α, RANTES, MCP-1; not MCP-2, MIP-1β or IL-8[1]

Expression pattern

Hematopoietic cells

CC CKR-4 mRNA found in IL-5-treated basophils[1]

Tissues

CC CKR-4 mRNA found in leukocyte-rich tissue[1]

Protein structure

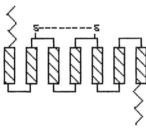

360 amino acids

Amino acid sequence

```
  1   MNPTDIADTT LDESIYSNYY LYESIPKPCT KEGIKAFGEL FLPPLYSLVF
 51   VFGLLGNSVV VLVLFKYKRL RSMTDVYLLN LAISDLLFVF SLPFWGYYAA
101   DQWVFGLGLC KMISWMYLVG FYSGIFFVML MSIDRYLAIV HAVFSLRART
151   LTYGVITSLA TWSVAVFASL PGFLFSTCYT ERNHTYCKTK YSLNSTTWKV
201   LSSLEINILG LVIPLGIMLF CYSMIIRTLQ HCKNEKKNKA VKMIFAVVVL
251   FLGFWTPYNI VLFLETLVEL EVLQDCTFER YLDYAIQATE TLAFVHCCLN
301   PIIYFFLGEK FRKYILQLFK TCRGLFVLCQ YCGLLQIYSA DTPSSSYTQS
351   TMDHDLHDAL
```

Database accession numbers

Source	EMBL/GenBank	Ref
Human	X85740	1

Reference
[1] Power, C.A. et al. (1995) *J. Biol. Chem.* 270, 19495–19500.

Chemoattractant
Receptors

Family

G-protein coupled receptor family, chemokine receptor branch of rhodopsin family

Tissue sources

U937 cells[1], HL-60 cells[2]

In vitro biological effects

COS-7 cells expressing C5aR turn over PI in response to C5a[1].

Ligands and ligand binding studies

- COS-7 cells expressing C5aR bind C5a with high affinity[2,3].
- HEK 293 cells expressing C5aR bind C5a with high affinity[4].
- C5a peptide C009 binds to HEK 293 cells expressing C5aR[4].

Expression pattern

Hematopoietic cells

C5aR mRNA is found in peripheral blood monocytes, peripheral blood granulocytes, and in the myeloid fraction of bone marrow[1].

Cell lines

C5aR message is found in the KG-1 monocytic cell line, retinoic acid-differentiated HL-60 cells, U937 cells stimulated with PMA or cAMP-induced U937 cells[1].

Gene structure

Murine

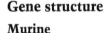

- Chromosomal location: Ch 19[5]
- Intronless coding region[2]; 2.2 kb mRNA

Protein structure

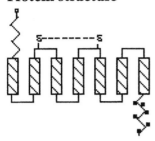

- 350 amino acids, calculated MW 39 320; 43 kD calculated from cross-linking data[3].
- Phosphorylation exclusively on serine residues in the C-terminal tail[6].
- Putative glycosylation sites: Asn-5.
- Putative disulfide bonds: Cys-109–Cys-188.
- Mutagenesis: N-terminal aspartic acid (D27) is important to ligand binding; N-terminus is not required for signaling[7] but is required for high-affinity binding to C5a[4]; Arg-40 and His-15 important for ligand binding[8]; Asp 282 is critical for ligand binding; Arg 206 is important for high affinity binding and is involved in G-protein coupling[9].

Amino acid sequence

Human

```
1    MNSFNYTTPD YGHYDDKDTL DLNTPVDKTS NTLRVPDILA LVIFAVVFLV
51   GVLGNALVVW VTAFEAKRTI NAIWFLNLAV ADFLSCLALP ILFTSIVQHH
101  HWPFGGAACS ILPSLILLNM YASILLLATI SADRFLLVFK PIWCQNFRGA
151  GLAWIACAVA WGLALLLTIP SFLYRVVREE YFPPKVLCGV DYSHDKRRER
201  AVAIVRLVLG FLWPLLTLTI CYTFILLRTW SRRATRSTKT LKVVVAVVAS
251  FFIFWLPYQV TGIMMSFLEP SSPTFLLLNK LDSLCVSFAY INCCINPIIY
301  VVAGQGFQGR LRKSLPSLLR NVLTEESVVR ESKSFTRSTV DTMAQKTQAV
```

Mouse

```
1    MNSSFEINYD HYGTMDPNIP ADGIHLPKRQ PGDVAALIIY SVVFLVGVPG
51   NALVVWVTAF EPDGPSNAIW FLNLAVADLL SCLAMPVLFT TVLNHNYWYF
101  DATACIVLPS LILLNMYASI LLLATISADR FLLVFKPIWC QKVRGTGLAW
151  MACGVAWVLA LLLTIPSFVY REAYKDFYSE HTVCGINYGG GSFPKEKAVA
201  ILRLMVGFVL PLLTLNICYT FLLLRTWSRK ATRSTKTLKV VMAVVICFFI
251  FWLPYQVTGV MIAWLPPSSP TLKRVEKLNS LCVSLAYINC CVNPIIYVMA
301  GQGFHGRLLR SLPSIIRNAL SEDSVGRDSK TFTPSTDDTS PRKSQAV
```

Dog

```
1    MASMNFSPPE YPDYGTATLD PNIFVDESLN TPKLSVPDMI ALVIFVMVFL
51   VGVPGNFLVV WVTGFEVRRT INAIWFLNLA VADLLSCLAL PILFSSIVQQ
101  GYWPFGNAAC RILPSLILLN MYASILLLTT ISADRFVLVF NPIWCQNYRG
151  PQLAWAACSV AWAVALLLTV PSFIFRGVHT EYFPFWMTCG VDYSGVGVLV
201  ERGVAILRLL MGFLGPLVIL SICYTFLLIR TWSRKATRST KTLKVVVAVV
251  VSFFVLWLPY QVTGMMMALF YKHSESFRRV SRLDSLCVAV AYINCCINPI
301  IYVLAAQGFH SRFLKSLPAR LRQVLAEESV GRDSKSITLS TVDTPAQKSQ
351  GV
```

Database accession numbers

Source	PIR	SwissProt	EMBL	Genbank	Ref
Human	A37963 S13646 S30518	P21730	X58674,57250	M62505	1
Mouse			L05630		10
Dog			X65860		11

References

[1] Gerard, N.P. and Gerard, C. (1991) *Nature* 349, 614–617.
[2] Boulay, F. et al. (1991) *Biochemistry* 30, 2993–2999.
[3] Gerard, N.P. and Gerard, C. (1990) *Biochemistry* 29, 9274–9281.
[4] Siciliano, P.A. et al. (1994) *Proc. Natl Acad. Sci. USA* 91, 1214–1218.
[5] Bao, L. et al. (1992) *Genomics* 13, 437–440.
[6] Raffetseder, D. et al. (1996) *Eur. J. Biochem.* 235, 82–90.
[7] DeMartino, J.A. et al. (1994) *J. Biol. Chem.* 269, 14446–14450.
[8] Mollison, K.W. et al. (1989) *Proc. Natl Acad. Sci. USA* 86, 292–296.
[9] Giannini, E. et al. (1995) *J. Biol. Chem.* 270, 19166–19172.
[10] Gerard, N.P. et al. (1992) *J. Immunol.* 149, 2600–2606.
[11] Perret, J.J. et al. (1992) *Biochem. J.* 288, 911–917.

FPR N-formyl peptide receptor

Alternate name

N-formylmethionyl-leucyl-phenylalanine receptor (FMLPR)

Family

G-protein coupled receptor family, chemokine receptor branch of rhodopsin family

Homologs

FPRH1 and FPRH2[2]

Tissue sources

HL-60 cells differentiated with Bt2 cAMP[1]

In vitro biological effects

- G-protein coupling[3], PT-sensitive[4].
- Calcium mobilization in HL-60 cells transfected with FPR ($EC_{50} = 3$ nM)[3].
- Inhibition of cAMP accumulation[5].
- PT-sensitive actin polymerization in HL-60 cells transfected with FPR ($EC_{50} = 10$ nM) but not in mouse L cells transfected with FPR[3].
- *Xenopus laevis* oocytes injected with FPR RNA show calcium mobilization in response to FMLP[6].

Ligands and ligand binding studies

- Undifferentiated HL-60 cells transfected with FPR bound FMLP with two affinities (0.6 and 33 nM)[3].
- Mouse L cells transfected with FPR bound FMLP with high affinity[3].
- COS7 cells transfected with FPR bound FMLPK-Pep12[1] with low and high affinity[10].

Expression pattern

Hematopoietic cells

1.6–1.7 kb FPR mRNA expressed in neutrophils.

Cell lines

1.6–1.7 kb FPR mRNA present in HL-60 cells[7].

Regulation of expression

Dibutyryl cAMP induces transcription of FPR in HL-60 cells[7].

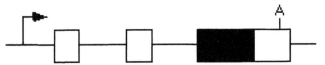

Gene structure

Human

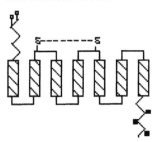

- Chromosomal location: 19q13.4[2].
- Exon/introns: entire coding region is found in exon 3[8].
- Alternative splicing of exon 2[7].

Protein structure

- 350 amino acids; calculated MW = 38 420[1].
- Phosphorylation sites: C-terminus phosphorylated by GPCR kinase 2.
- Putative glycosylation sites: Asn-4, Asn-10[2].
- Mutagenesis: D71A, R123G, R309G/E310A/R311G are all uncoupled from the G protein, probably by their incapacity to form a high affinity ligand receptor complex[9].

Amino acid sequence

Human FMLPR

```
1    METNSSLPTN ISGGTPAVSA GYLFLDIITY LVFAVTFVLG VLGNGLVIWV
51   AGFRMTHTVT TISYLNLAVA DFCFTSTLPF FMVRKAMGGH WPFGWFLCKF
101  VFTIVDINLF GSVFLIALIA LDRCVCVLHP VWTQNHRTVS LAKKVIIGPW
151  VMALLLTLPV IIRVTTVPGK TGTVACTFNF SPWTNDPKER IKVAVAMLTV
201  RGIIRFIIGF SAPMSIVAVS YGLIATKIHK QGLIKSSRPL RVLSFVAAAF
251  FLCWSPYQVV ALIATVRIRE LLQGMYKEIG IAVDVTSALA FFNSCLNPML
301  YVFMGQDFRE RLIHALPASL ERALTEDSTQ TSDTATNSTL PSAEVALQAK
```

Rabbit FMLPR

```
1    MDSNASLPLN VSGGTQATPA GLVVLDVFSY LILVVTFVLG VLGNGLVIWV
51   TGFRMTHTVT TISYLNLALA DFSFTSTLPF FIVTKALGGH WPFGWFLCKF
101  VFTIVDINLF GSVFLIALIA LDRCICVLHP VWAQNHRNVS LAKKVIVGPW
151  ICALLLTLPV IIRVTTLSHP RAPGKMACTF DWSPWTEDPA EKLKVAISMF
201  MVRGIIRFII GFSTPMSIVA VCYGLIATKI HRQGLIKSSR PLRVLSFVVA
251  SFLLCWSPYQ IAALIATVRI RELLLGMGKD LRIVLDVTSF VAFFNSCLNP
301  MLYVFMGQDF RERLIHSLPA SLERALSEDS AQTSDTGTNS TSAPAEAELQ
351  AI
```

Database accession numbers

	PIR	SwissProt	EMBL/GenBank	Ref
Human	A42009			10
Rabbit	A46520	Q05394	M94549	11

References

[1] Boulay, F. et al. (1990) *Biochemistry* 29, 11123–11133.

[2] Bao, L. et al. (1992) *Genomics* 13, 437–440.

[3] Prossnitz, E.R. et al. (1993) *J. Immunol.* 151, 5704–5715.

[4] Tsu, R.C. et al. (1995) *Biochem. J.* 309, 331–339.

[5] Lang, J. et al. (1993) *EMBO J.* 12, 2671–2679.

[6] Thomas, K.M. et al. (1990) *J. Biol. Chem.* 265, 20061–20064.

[7] Perez, H.D. et al. (1992) *J. Biol.Chem.* 267, 358–363.

[8] Murphy, P.M. et al. (1993) . *Gene* 133, 285–290.

[9] Prossnitz, E.R. et al. (1995) *J. Biol. Chem.* 270, 10686–10694.

[10] Boulay, F. et al. (1990) *Biochem. Biophys. Res. Commun.* 168, 1103–1109.

[11] Ye, R.D. et al. (1993) *J. Immunol.* 150, 1383–1394.

Miscellaneous
Receptors

Family

G-protein coupled receptor family, chemokine receptor branch of rhodopsin family

Tissue sources

HL-60 granulocyte cDNA library[1], U937[2]

In vitro biological effects

- Calcium mobilization in response to PAF in PAFR stably transfected fibro-blasts[1] and in PAFR-injected *Xenopus* oocytes[3].
- *Xenopus* oocytes injected with guinea pig PAFR or rat PAFR show calcium mobilization in response to PAF[4].
- 1,4,5-trisphosphate production is seen in response to PAF in PAFR-injected *Xenopus* oocytes and PAFR-transfected COS-7 cells[3].
- COS-7 cells expressing PAFR undergo ligand internalization[5].

Ligands and ligand binding studies

- The classic ligand is the proinflammatory lipid, PAF.
- PAFR antagonists include L-659,989, WEB 2086[1], SRI-63675[6], carbarmyl-PAF (non-metabolizable PAF analog, $EC_{50} = 10\,nM$)[6], WEB 2170[7].
- COS-7 cells expressing PAFR have a single high-affinity binding site for PAF[3,8].
- Undifferentiated HL-60 cells transfected with PAFR bind WEB 2086 ($K_d = 30.7\,nM$)[1].

Expression pattern

Hematopoietic cells

Peripheral leukocytes[3] (transcript #1[9]) – see Gene Structure

Tissues

Transcript #1 is ubiquitous[9]; transcript #2 in placenta, heart and lung[1]

Cell lines

Differentiated HL-60[1], U937[6]; KS cells[7]; differentiated eosinophil-like cell line EoL-1[3,9]

Regulation of expression

DMSO and retinoic acid in HL-60 cells[10] upregulate PAFR expression.
PAFR mRNA transcription is down-regulated in U937 cells by carbarmyl-PAF[6].
GM-CSF, IL-5, and *n*-butyrate increases PAFR mRNA expression in EoL-1 cells[3].
Retinoic acid and 3,3′,5-triiodothyronine increase transcript #2 expression in heart and skin. Vitamin D had no effect[13].

Gene structure

Human

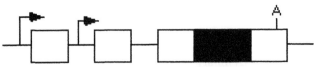

- Chromosomal location: chromosome 1[9,11].
- Exon/introns: two noncoding exons, Ex1 and Ex2[9], alternative splicing produces transcript #1 (Exons 1 and 3) and transcript #2 (Exons 2 and 3)[9].
- Promoter: TATA-less, initiation at −137[10], promoter upstream of exon 1 with NF-κB and SP-1 binding sites and second promoter upstream of exon 2 with AP-1, AP-2, and SP-1 binding sites[9].
- 4.0 kb mRNA[1].
- Promoter 2 contains a 24 basepair hormone response element[13].

Protein structure

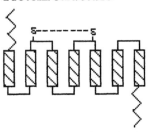

- 342 amino acids, calculated MW 39 203.
- Putative phosphorylation sites: not known.
- Putative glycosylation sites: no N-terminal sites[1].
- Putative disulfide bonds: 90–173[1].

Amino acid sequences

Human

```
1    MEPHDSSHMD SEFRYTLFPI VYSIIFVLGV IANGYVLWVF ARLYPCKKFN
51   EIKIFMVNLT MADMLFLITL PLWIVYYQNQ GNWILPKFLC NVAGCLFFIN
101  TYCSVAFLGV ITYNRFQAVT RPIKTAQANT RKRGISLSLV IWVAIVGAAS
151  YFLILDSTNT VPDSAGSGNV TRCFEHYEKG SVPVLIIHIF IVFSFFLVFL
201  IILFCNLVII RTLLMQPVQQ QRNAEVKRRA LWMVCTVLAV FIICFVPHHV
251  VQLPWTLAEL GFQDSKFHQA INDAHQVTLC LLSTNCVLDP VIYCFLTKKF
301  RKHLTEKFYS MRSSRKCSRA TTDTVTEVVV PFNQIPGNSL KN
```

Rat

```
1    MEQNGSFRVD SEFRYTLFPI VYSVIFVLGV VANGYVLWVF ATLYPSKKLN
51   EIKIFMVNLT VADLLFLMTL PLWIVYYSNE GDWIVHKFLC NLAGCLFFIN
101  TYCSVAFLGV ITYNRYQAVA YPIKTAQATT RKRGITLSLV IWISIAATAS
151  YFLATDSTNV VPKKDGSGNI TRCFEHYEPY SVPILVVHIF ITSCFFLVFF
201  LIFYCNMVII HTLLTRPVRQ QRKPEVKRRA LWMVCTVLAV FVICFVPHHV
251  VQLPWTLAEL GYQTNFHQAI NDAHQITLCL LSTNCVLDPV IYCFLTKKFR
301  KHLSEKFYSM RSSRKCSRAT SDTCTEVMMP ANQTPVLPLK N
```

Guinea pig

```
1   MELNSSSRVD SEFRYTLFPI VYSIIFVLGI IANGYVLWVF ARLYPSKKLN
51  EIKIFMVNLT VADLLFLITL PLWIVYYSNQ GNWFLPKFLC NLAGCLFFIN
101 TYCSVAFLGV ITYNRFQAVK YPIKTAQATT RKRGIALSLV IWVAIVAAAS
151 YFLVMDSTNV VSNKAGSGNI TRCFEHYEKG SKPVLIIHIC IVLGFFIVFL
201 LILFCNLVII HTLLRQPVKQ QRNAEVRRRA LWMVCTVLAV FVICFVPHHM
251 VQLPWTLAEL GMWPSSNHQA INDAHQVTLC LLSTNCVLDP VIYCFLTKKF
301 RKHLSEKLNI MRSSQKCSRV TTDTGTEMAI PINHTPVNPI KN
```

Rhesus monkey

```
1   MEPHDSSHVD SEFRYTLFPI VYSIIFVLGV IANGYVLWVF ARLYPSKKFN
51  EIKIFMVNLT MADMLFLITL PLWIVYYQNG GNWIFPKFLC NLAGCLFFIN
101 TYCSVAFLGV ITYNRFQAVT RPIKTAQANT RKRGISLSLV IWVAIVGAAS
151 YFFILDSTNT VPNSAGSGNI TRCFEHYEKG SVPVLIIHIF IVFSFFLVFL
201 IILFCNLV
```

Database accession numbers

	PIR	SwissProt	EMBL/GenBank	Ref
Human	JH0479	P25105	M80436	1
	A40191		M76674	8
	A41079		A42831	11
			M88177	
			L07334	12
Rat	S43252		U04740	4
Guinea pig		P21556	X56736	5
Rhesus monkey			L07333	12

References

1 Ye, R.D. et al. (1991) *Biochem. Biophys. Res. Commun.* 180, 105–111.
2 Gerard, N. and Gerard, C. (1994) *J. Lipid Mediators Cell Signalling* 10, 77–80.
3 Nakamura, M. et al. (1991) *J. Biol. Chem.* Vol. 20400–20405.
4 Bito, H. et al. (1994) *Eur. J. Biochem.* 221, 211–218
5 Honda, Z. et al. (1991) *Nature* 349, 342–346.
6 Chau, L.Y. et al. (1994) *Biochem. J.* 301, 911–916.
7 Bussolino, F. et al. (1995) *J. Clin. Invest.* 96, 940–952.
8 Kunz, D. et al. (1992) *J. Biol. Chem.* 267, 9101–9106.
9 Mutoh, H. et al. (1993) *FEBS Lett.* 322, 129–134.
10 Pang, J.H. et al. (1995) *J. Biol. Chem.* 270, 14123–14129.
11 Seyfried, C.E. et al. (1992) *Genomics* 13, 832–834.
12 Behal, R.H. et al. (1992) direct submission.
13 Mutoh, H. et al. (1996) *Proc. Natl. Acad. Sci.* 93, 774–779.

 **DARC** Duffy antigen/chemokine receptor

Alternate names

Erythrocyte chemokine receptor (erythrocyte CKR), RBC chemokine receptor, multispecific chemokine receptor, gpFy, gpD

Family

G-protein coupled receptor family, chemokine receptor branch of the rhodopsin family

Tissue sources

Human bone marrow[1], HEL cells[2]

In vitro biological effects

- K562 cells transfected with DARC internalize radiolabeled MGSA[3].
- No regulation by G proteins[4].

Ligands and ligand binding studies

- Human malarial parasite *Plasmodium vivax*.
- β-chemokines: RANTES, MCP-1, not MIP-1α or β[2].
- α-chemokines: IL-8 ($K_d = 9.5\,\text{nM} \pm 3.6$), MGSA/GRO.
- K562 cells transfected with DARC bind MGSA ($K_d = 3\,\text{nM}$)[3].
- HEK 293 cells transfected with DARC bind RANTES, MCP-1, IL-8 and MGSA/GRO with high affinity[4].
- ^{125}I IL-8 bound to DARC is displaced by IL-8 ($K_d = 9.5 \pm 3.6\,\text{nM}$) and all chemokines but MIP-1α[2].

Expression pattern
Hematopoietic cells

Expressed in erythroid cells, bone marrow[1]

Tissues

Expressed in endothelial cells lining postcapillary venules[6] and splenic sinusoids of Duffy negative individuals[3], adult spleen, kidney, brain, fetal liver[1]

Cell lines

Expressed in K562[1,] HEL[2]

Gene structure

- Exon/introns: single exon[7,8].
- Alternative splicing: alternative polyA sites[8]; 1.27 kb, 2.2 kb, 8.5 kb (brain) mRNAs[1].
- Promoter: GATA1 site at −46 disrupted by a T → C change that leads to repression of DARC expression in carriers of the silent FY*B allele[7]; AP-1, HNF-5, TCF-1, ApoE B[2], W-element, H-APD-1, Sp-1; no TATA or CCAAT boxes[8].

Protein structure

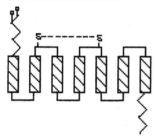

- 338 amino acids; 40 kD in erythrocyte ghosts and 50 kD in HEL cells[2], 43–45 kD renal isoform[6]; 47 kD in K562 cells transfected with DARC[5].
- Putative glycosylation sites: Asn-18, Asn-29[1]; glycanase treatment causes a shift in the size of the receptor from 47 to 42 kD[2].
- Mutagenesis studies: N-terminal domain of DARC is required for its promiscuous binding, since a chimeric receptor in which the N-terminal domain of DARC on an IL-8RB body bound both α-chemokines IL-8 and MGSA as well as β-chemokines RANTES and MCP-1, similar to wild-type DARC[9].
- Naturally occurring mutations: A → G change at base 131 results in amino acid change of Gly-44 → Asp in the N-terminal domain. This change produces the antigens Fya or Fyb, respectively[10].

Amino acid sequence

```
1   MASSGYVLQA ELSPSTENSS QLDFEDVWNS SYGVNDSFPD GDYDANLEAA
51  APCHSCNLLD DSALPFFILT SVLGILASST VLFMLFRPLF RWQLCPGWPV
101 LAQLAVGSAL FSIVVPVLAP GLGSTRSSAL CSLGYCVWYG SAFAQALLLG
151 CHASLGHRLG AGQVPGLTLG LTVGIWGVAA LLTLPVTLAS GASGGLCTLI
201 YSTELKALQA THTVACLAIF VLLPLGLFGA KGLKKALGMG PGPWMNILWA
251 WFIFWWPHGV VLGLDFLVRS KLLLLSTCLA QQALDLLLNL AEALAILHCV
301 ATPLLLALFC HQATRTLLPS LPLPEGWSSH LDTLGSKS
```

Database accession numbers

Source	GenBank	Ref
Human	U01839	1,2

References

[1] Chaudhuri, A. et al. (1993) *Proc. Natl. Acad. Sci. USA* 90, 10793–10797.

[2] Horuk, R. et al. (1994) *J. Biol. Chem.* 269, 17730–17733.

[3] Peiper, S. (1995) *J. Exp. Med.* 181, 1311–1317.

[4] Neote, K. et al. (1994) *Blood* 84, 44–52.

[5] Chaudhuri, A. et al. (1994) *J. Biol. Chem.* 269, 7835–7838.

[6] Hadley, T.J. et al. (1994) *J. Clin. Invest.* 94, 985–991.

[7] Tournamille, C. et al. (1995) *Nat. Genet.* 10, 224–228.

[8] Iwamoto, S. et al. (1995) *Blood* 85, 622–626.

[9] Lu, Z.H. et al. (1995) *J. Biol. Chem.* 270, 26239–26245.

[10] Mallinson, G. et al. (1995) *Br. J. Haematol.* 90, 823–829.

Viral
Chemokine
Receptors

Alternate names

Human cytomegalovirus (CMV) G-protein coupled receptor

Family

G-protein coupled receptor family, chemokine receptor branch of the rhodopsin family

Homologs

UL33, US27[1], capripoxvirus Q2/3L[3]

Tissue sources

Human CMV DNA[1]; CMV-infected human fibroblasts[2]

In vitro biological effects

Calcium mobilization in US28-transfected K562 cells in response to RANTES ($EC_{50} = 5$ nM)[2].

Ligands and ligand binding studies

- COS-7 cells transfected with US28 bound ^{125}I-MCP-1 ($K_d = 6.0 \times 10^{-10}$ M); ^{125}I-RANTES ($K_d = 2.7 \times 10^{-10}$ M); and these could be competed by unlabeled MIP-1α and MIP-1β ($K_d = 1.2$–7.5×10^{-9} M)[4].
- 293 cells transiently transfected with US28 bind MIP-1α ($K_d = 5$ nM)[5].
- K562 cells stably transfected with US28 bind MIP-1α ($K_d = 2$–6 nM) and this binding can be completed by MCP-1, MIP-1α, MIP-1β, and RANTES[2].

Expression pattern

Hematopoietic cells

Late phase of lytic infection of leukocytes

Tissues

Late phase of lytic infection of epithelial cells, fibroblasts, smooth muscle cells.

Gene structure

- Message size: 1.5 kb[2]
- Chromosomal location: N/A (viral)

Protein structure

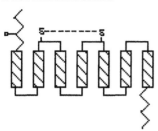

- 354 amino acids
- Putative glycosylation sites: Asn-30

Amino acid sequence

```
1   MTPTTTTAEL TTEFDYDEDA TPCVFTDVLN QSKPVTLFLY GVVFLFGSIG
51  NFLVIFTITW RRRIQCSGDV YFINLAAADL LFVCTLPLWM QYLLDHNSLA
101 SVPCTLLTAC FYVAMFASLC FITEIALDRY YAIVYMRYRP VKQACLFSIF
151 WWIFAVIIAI PHFMVVTKKD NQCMTDYDYL EVSYPIILNV ELMLGAFVIP
201 LSVISYCYYR ISRIVAVSQS RHKGRIVRVL IAVVLVFIIF WLPYHLTLFV
251 DTLKLLKWIS SSCEFERSLK RALILTESLA FCHCCLNPLL YVFVGTKFRQ
301 ELHCLLAEFR QRLFSRDVSW YHSMSFSRRS SPSRRETSSD TLSDEVCRVS
351 QIIP
```

Database accession numbers

Source	SwissProt	GenBank	Ref
Human	P32952	L20501	5
		X17403	6

References

[1] Chee, M.S. et al. (1990) *Nature* 344, 774–777.
[2] Gao, J.L. and Murphy, P.M. (1994) *J. Biol. Chem.* 269, 28539–28542.
[3] Cao, J.X. et al. (1995) *Virology* 209, 207–212.
[4] Kuhn, D.E. et al. (1995) *Biochem. Biophys. Res. Commun.* 221, 325–330.
[5] Neote, K. et al. (1993) *Cell* 72, 415–425.
[6] Waldman, W. J. et al. (1991) *Arch. Virol.* 117, 143–164.

ECRF3

ECRF3

Family

G-protein coupled receptor family, chemokine receptor branch of the rhodopsin family

Tissue sources

Herpesvirus saimiri of the γ-herpesvirus family[1]

In vitro biological effects

Calcium efflux

Ligands and ligand binding studies

MGSA/GRO-α > NAP-2 > IL-8[1]

Expression pattern

Hematopoietic cells

T cells

Protein structure

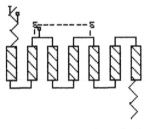

- 321 amino acids, calculated MW = 37 090
- Putative glycosylation sites: Asn-14, Asn-18, Asn-117

Amino acid sequence

```
1    MEVKLDFSSE DFSNYSYNYS GDIYYGDVAP CVVNFLISES ALAFIYVLMF
51   LCNAIGNSLV LRTFLKYRAQ AQSFDYLMMG FCLNSLFLAG YLLMRLLRMF
101  EIFMNTELCK LEAFFLNLSI YWSPFILVFI SVLRCLLIFC ATRLWVKKTL
151  IGQVFLCCSF VLACFGALPH VMVTSYYEPS SCIEEDGVLT EQLRTKLNTF
201  HTWYSFAGPL FITVICYSMS CYKLFKTKLS KRAEVVTIIT MTTLLFIVFC
251  IPYYIMESID TLLRVGVIEE TCAKRSAIVY GIQCTYMLLV LYYCMLPLMF
301  AMFGSLFRQR MAAWCKTICH C
```

Database accession numbers

	PIR	SwissProt	GenBank	EMBL	Ref
Human	S20245	Q01035	M86409	X64346	1

Reference

[1] Ahuja, S.K. and Murphy, P.M. (1993) *J. Biol. Chem.* 268, 20691–20694.

Orphan
Chemokine
Receptors

LESTR — Leukocyte-derived seven transmembrane domain receptor

Alternate names

FB22, NPYRL, HUMSTR[4], fusin[5], pBE 1.3[4], CXCR4[5]

Family

G-protein coupled receptor family, chemokine receptor branch of rhodopsin family

Homologs

LCR1[1] (bovine)

Tissue sources

Human blood monocyte cDNA library[2], human fetal spleen[4]

Ligands and ligand binding studies[a]

CHO and COS cells transfected with LESTR do not bind IL-8, NAP-2, GRO-α, MCP-1, MCP-3, MIP-1α, RANTES[2]

Expression pattern
Hematopoietic cells

Monocytes, neutrophils, lymphocytes, PHA-activated T cell blasts[2]

Cell lines

Vitamin D_3-treated HL-60; U937, Jurkat, DMSO-treated HL-60[2]

Gene structure

- 1.7–1.8 kD mRNA[2]
- Locus D25201E is located on human chromosone 2q21[4]

Protein structure

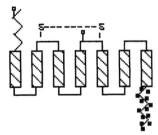

[a] Recently, LESTR/fusin has been shown to be the receptor for SDF-1[6,7] and play a role in HIV-infection of T cells[7]. Thus, it is no longer an orphan receptor, but a member of the CXC or β receptor family, and has been renamed CXCR4.

- 352 amino acids; calculated MW = 39 745[2]
- Putative phosphorylation sites: 18 serine and threonine residues in C-terminus
- Putative glycosylation sites: Asn-11[3], Asn-177[2]
- Putative disulfide bonds: Cys-109–Cys-186[3]

Amino acid sequence

```
1    MEGISIYTSD NYTEEMGSGD YDSMKEPCFR EENANFNKIF LPTIYSIIFL
51   TGIVGNGLVI LVMGYQKKLR SMTDKYRLHL SVADLLFVIT LPFWAVDAVA
101  NWYFGNFLCK AVHVIYTVNL YSSVLILAFI SLDRYLAIVH ATNSQRPRKL
151  LAEKVVYVGV WIPALLLTIP DFIFANVSEA DDRYICDRFY PNDLWVVVFQ
201  FQHIMVGLIL PGIVILSCYC IIISKLSHSK GHQKRKALKT TVILILAFFA
251  CWLPYYIGIS IDSFILLEII KQGCEFENTV HKWISITEAL AFFHCCLNPI
301  LYAFLGAKFK TSAQHALTSV SRGSSLKILS KGKRGGHSSV STESESSSFH
351  SS
```

Database accession numbers

Source	PIR	SwissProt	GenBank	EMBL	Ref
Human	S32761	P30991	L06797	X71635	3
	A45747		M99293		4

References
[1] Peveri, P. et al. (1988) *J. Exp. Med.* 167, 1547–1559.
[2] Loetscher, M. et al. (1994) *J. Biol. Chem.* 269, 232–237.
[3] Herzog, H. et al. (1993) *DNA Cell Biol.* 12, 465–471.
[4] Federsppiel, B. et al. (1993) *Genomics* 16, 707–712.
[5] Feng, Y. et al. (1996) *Science* 272, 872–877.
[6] Bluel, C.C. et al. (1996) *Nature* 382, 829–833.
[7] Oberlin, E. et al. (1996) *Nature* 382, 833–835.

CMK-BRL-1 · Chemokine β receptor-like 1

Family

G-protein coupled receptor family, chemokine receptor branch of the rhodopsin family

Homologs

- 40% identical to β-chemokine receptors[1]
- 83% identical to rat orphan receptor, RBS11[2]

Tissue sources

Eosinophilic leukemia library

Expression pattern

Hematopoietic cells

Neutrophils, monocytes[1], not eosinophils[1]

Tissues

Brain, placenta, lung, liver, pancreas[1]

Gene structure

- Chromosomal location 3p21.
- Exon/introns: 1116 bp ORF, exon 1 encodes 17 amino acids followed by a 9 kb intron and the remaining coding region in Exon 2.

Protein structure

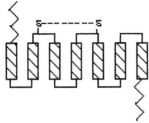

372 amino acids; MW = 40 369 Daltons

Amino acid sequence

Human

```
1    MDQFPESVTE NFEYDDLAEA CYIGDIVVFG TVFLSIFYSV IFAIGLVGNL LVVFALTNSK
51   KPKSVTDIYL LNLALSDLLF VATLPFWTHY LINEKGLHNA MCKFTTAFFF IGFFGSIFFI
101  TVISIDRYLA IVLAANSMNN RTVQHGVTIS LGVWAAAILV AAPQFMFTKQ KENECLGDYP
151  EVLQEIWPVL RNVETNFLGF LLPLLIMSYC YFRIIQTLFS CKNHKKAKAI KLILLVVIVF
201  FLFWTPYNVM IFLETLKLYD FFPSCDMRKD LRLALSVTET VAFSHCCLNP LIYAFAGEKF
251  RRYLYHLYGK CLAVLCGRSV HVDFSSSESQ RSRHGSVLSS NFTYHTSDGD ALLLL
```

Database accession numbers

Source	PIR	EMBL	Ref
Human		U28934	1

References
[1] Combadiere, C. et al. (1995) *DNA Cell Biol.* 14, 673–680.
[2] Harrison, J.K. (1994) *Neurosci. Lett.* 169, 85–89.

BLR-1/MDR-15 — Burkitt's lymphoma receptor-1/ Monocyte-derived receptor-15

Alternate names

BLR-1[1], MDR-15[2]

Family

G-protein coupled receptor family, chemokine receptor branch of the rhodopsin family

Homologs

- 40% amino acid identity to IL-8 receptors
- mBLR1 (murine homolog)[3]

Tissue sources

Burkitt's lymphoma cell line BL64[1]

Ligands and ligand binding studies

- Jurkat cells expressing BLR1 do not bind known chemokines[2].
- CHO and COS cells expressing BLR1 do not bind known chemokines[1].

Expression pattern

Hematopoietic cells

- B cells and memory T cells[3].
- Chronic B-lymphoid leukemia and non-Hodgkin's lymphoma cells, peripheral blood monocytes, and lymphocytes.

Tissues

- Expression in tonsils and secondary lymphatic organs[1].
- Granule and Purkinje cell layer of the cerebellum.
- Expressed in fetal liver and brain during late embryogenesis[4].

Cell lines

Most Burkitt's lymphoma cell lines tested[1]

Regulation of expression

- IL-4 and IL-6 down-regulate BLR-1[2,3]
- CD40 and CD3 MoAb led to down-regulation of BLR-1[5]

Gene structure

Human

- Gene structure: two exons; 3.1 kb BLR1 mRNA.
- Alternative splicing: The MDR15 ORF is 45 codons shorter than BLR1 at the 5′ end.

Protein structure

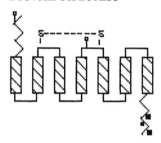

- 327 amino acids (MDR-15) or 371 amino acids (BLR-1)
- Putative phosphorylation sites: Ser-354, Ser-359, Ser-361
- Putative glycosylation sites: Asn-26, Asn-196
- Putative disulfide bridges: Csy-122–Csy-202

Amino acid sequence

MDR-15

```
1   MASFKAVFVP VAYSLIFLLG VIGNVLVLVI LERHRQTRSS TETFLFHLAV
51  ADLLLVFILP FAVAEGSVGW VLGTFLCKTV IALHKVNFYC SSLLLACIAV
101 DRYLAIVHAV HAYRHRRLLS IHITCGTIWL VGFLLALPEI LFAKVSQGHH
151 NNSLPRCTFS QENQAETHAW FTSRFLYHVA GFLLPMLVMG WCYVGVVHRL
201 RQAQRRPQRQ KAVRVAILVT SIFFLCWSPY HIVIFLDTLA RLKAVDNTCK
251 LNGSLPVAIT MCEFLGLAHC CLNPMLYTFA GVKFRSDLSR LLTKLGCTSP
301 ASLCQLFPSW RRSSLSESEN ATSLTTF
```

BLR-1

```
1   MNYPLTLEMD LENLEDLFWE LDRLDNYNDT SLVENHLCPA TEGPLMASFK
51  AVFVPVAYSL IFLLGVIGNV LVLVILERHR QTRSSTETFL FHLAVADLLL
101 VFILPFAVAE GSVGWVLGTF LCKTVIALHK VNFYCSSLLL ACIAVDRYLA
151 IVHAVHAYRH RRLLSIHITC GTIWLVGFLL ALPEILFAKV SQGHHNNSLP
201 RCTFSQENQA ETHAWFTSRF LYHVAGFLLP MLVMGWCYVG VVHRLRQAQR
251 RPQRQKAVRV AILVTSIFFL CWSPYHIVIF LDTLARLKAV DNTCKLNGSL
301 PVAITMCEFL GLAHCCLNPM LYTFAGVKFR SDLSRLLTKL GCTGPASCCQ
351 LFPSWRRSSL SESENATSLT TF
```

Database accession numbers

Source	GenBank	Ref
Human	X68129	2 (MDR-15)
	X68149	1 (BLR-1)

References
[1] Dobner, T. et al. (1992) *Eur. J. Immunol* 22, 2795–2799.
[2] Barella, L. et al. (1995) *Biochem. J.* 309, 773–779.
[3] Forster, R. et al. (1994) *Cell Mol. Biol* 40, 381–387.
[4] Kaiser, X. et al. (1993) *Eur. J. Immunol* 23, 2532–2539.
[5] Forster, R. et al. (1994) *Blood* 84, 830–840.

Tissue sources

Peripheral blood mononuclear cells[1]

Expression pattern

Hematopoietic cells

Expressed in neural and lymphoid tissue[1]

Cell lines

THP-1 cells[1]

Gene structure

Human chromosome 3p21–3pter[1]

Amino acid sequence

```
1    MDQFPESVTE NFEYDDLAEA CYIGDIVVFG TVFLSIFYSV IFAIGLVGNL
51   LVVFALTNSK KPKSVTDIYL LNLALSDLLF VATLPFWTHY LINEKGLHNA
101  IGFFGSIFFI MCKFTTAFFF TVISIDRYLA IVLAANSMNN RTVQHGVTIS
151  LGVWAAAILV AAPQFMFTKQ KENECLGDYP EVLQEIWPVL RNVETNFLGF
201  LLPLLIMSYC YFRIIQTLFS CKNHKKAKAI KLILLVVIVF FLFWTPYNVM
251  IFLETLKLYD FFPSCDMRKD LRLALSVTET VAFSHCCLNP LIYAFAGEKF
301  RRYLYHLYGK CLAVLCGRSV HVDFSSSESQ RSRHGSVLSS NFTYHTSDGD
351  ALLLL
```

Database accession numbers

Source	EMBL/GenBank	Ref
Human	U20350	1

Reference

[1] Raport, C.J. et al. (1995) *Gene* 163, 295–299.

EBI1

EBI1

Alternate names

BLR2[1]

Tissue sources

Epstein–Barr-induced cDNA

Expression pattern
Hematopoietic cells

Normal lymphoid tissues

Cell lines

Several B and T lymphocyte cell lines, including BL64, Raji and ES111[1]

Regulation of expression

- EBI1 mRNA induced in PHA⁻ and anti-CD3-treated peripheral blood lymphocytes[1]
- EBI1 mRNA detected in EBV-positive B cell lines[1]
- EBI1 mRNA induced in EBV-negative BL4I cells by EBV-nuclear antigen 2[1]

Gene structure

- Human chromosome 17q12–21.2.
- Its first extracellular domain is interrupted by flanking introns.

Amino acid sequence
Human

```
1    MDLGKPMKSV LVVALLVIFQ VCLCQDEVTD DYIGDNTTVD YTLFESLCSK
51   KDVRNFKAWF LPIMYSIICF VGLLGNGLVV LTYIYFKRLK TMTDTYLLNL
101  AVADILFLLT LPFWAYSAAK SWVFGVHFCK LIFAIYKMSF FSGMLLLLCI
151  SIDRYVAIVQ AVSAHRHRAR VLLISKLSCV GIWILATVLS IPELLYSDLQ
201  RSSSEQAMRC SLITEHVEAF ITIQVAQMVI GFLVPLLAMS FCYLVIIRTL
251  LQARNFERNK AIKVIIAVVV VFIVFQLPYN GVVLAQTVAN FNITSSTCEL
301  SKQLNIAYDV TYSLACVRCC VNPFLYAFIG VKFRNDLFKL FKDLGCLSQE
351  QLRQWSSCRH IRRSSMSVEA ETTTTFSP
```

Mouse

```
1   MDPGKPRKNV LVVALLVIFQ VCFCQDEVTD DYIGENTTVD YTLYESVCFK
51  KDVRNFKAWF LPLMYSVICF VGLLGNGLVI LTYIYFKRLK TMTDTYLLNL
101 AVADILFLLI LPFWAYSEAK SWIFGVYLCK GIFGIYKLSF FSGMLLLLCI
151 SIDRYVAIVQ AVSRHRHRAR VLLISKLSCV GIWMLALFLS IPELLYSGLQ
201 KNSGEDTLRC SLVSAQVEAL ITIQVAQMVF GFLVPMLAMS FCYLIIIRTL
251 LQARNFERNK AIKVIIAVVV VFIVFQLPYN GVVLAQTVAN FNITNSSCET
301 SKQLNIAYDV TYSLASVRCC VNPFLYAFIG VKFRSDLFKL FKDLGCLSQE
351 RLRHWSSCRH VRNASVSMEA ETTTTFSP
```

Database accession numbers

Source	EMBL	GenBank	Ref
Human	L31582		2
Mouse	L31580, L31501		2

Reference
[1] Burgstahler, R. et al. (1995) *Biochem. Biophys. Res. Comm.* 215, 737–743.
[2] Schweickart, V.L. et al. (1994) *Genomics* 23, 643–650.

Index

Printed and bound by CPI Group (UK) Ltd, Croydon, CR0 4YY

03/10/2024

01040421-0020